Name _____   Skill: On the hour

What time is it?

1. ____   2. ____   3. ____   4. ____

5. ____   6. ____   7. ____   8. ____

9. ____   10. ____   11. ____   12. ____

13. ____   14. ____

© Frank Schaffer Publications, Inc.   FS-2674 Math Workbook–Time and Money

Name _____   Skill: On the hour

"Here's your new puppy, Andy," said Mrs. Nelson. "I hope you will take good care of him. Let me tell you what to do." Mrs. Nelson wrote a list. Andy took it home. He hung it on the wall.

Read the time words. Draw the hands on Andy's clocks.

Morning   ### Afternoon

a. seven   b. nine   c. ten   d. four

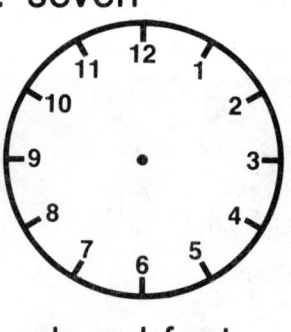 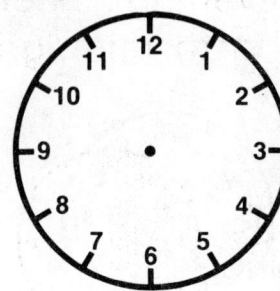 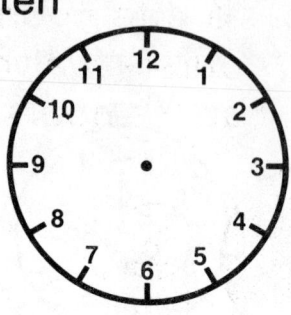 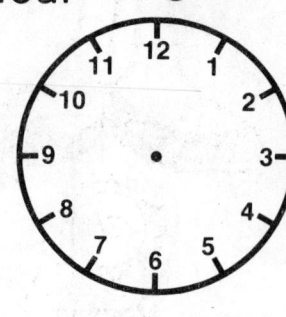

breakfast   bath   walk   brush fur

=============== Evening ===============

e. six   f. seven   g. nine   h. eleven

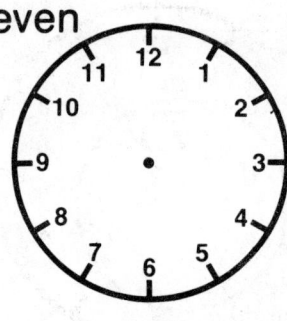 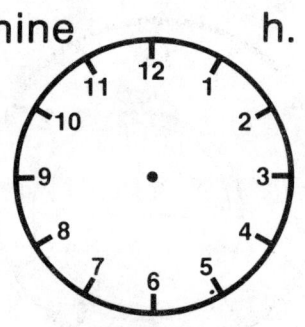 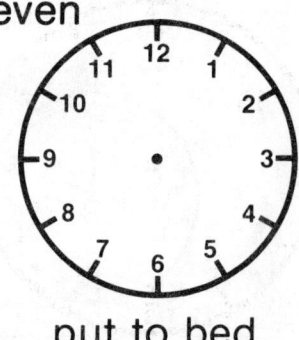

dinner   walk   read a story   put to bed

Pretend you have a pet gorilla. Write four things you would do for your gorilla. Draw the hands to show when you will do each of these things.

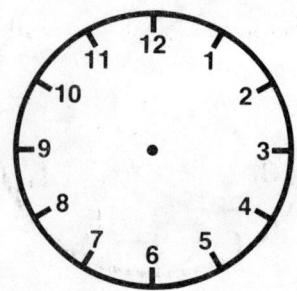 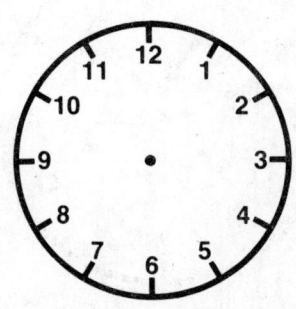 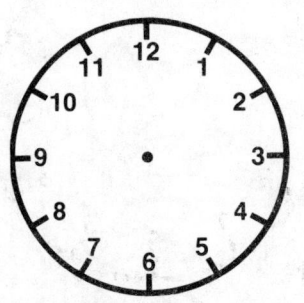 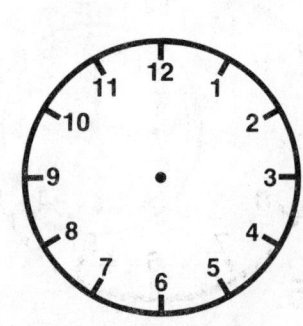

1. _____   2. _____   3. _____   4. _____

Name _____   Skill: AM and PM

"The show starts at nine o'clock," Ann told her mother. "Is that nine in the morning?"

"Yes, it is," answered Mother. "Do you know how to tell? Look at the letters after nine. It says a.m. That means 'in the morning.' There are twelve morning hours—midnight to twelve noon.

"Another show starts at eight p.m. P.M. means 'in the afternoon or evening'. There are twelve p.m. hours—twelve noon to midnight. Each day has 24 hours."

On the line, write the <u>hour</u> that Ann might do each thing. Write a.m. or p.m. Draw the hands on the clock.

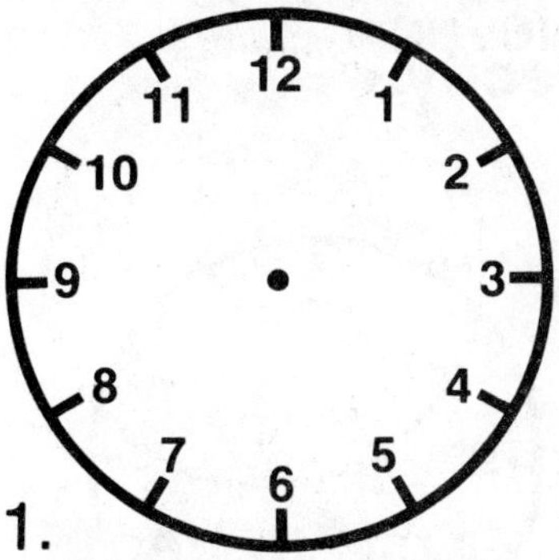

1. get home from school 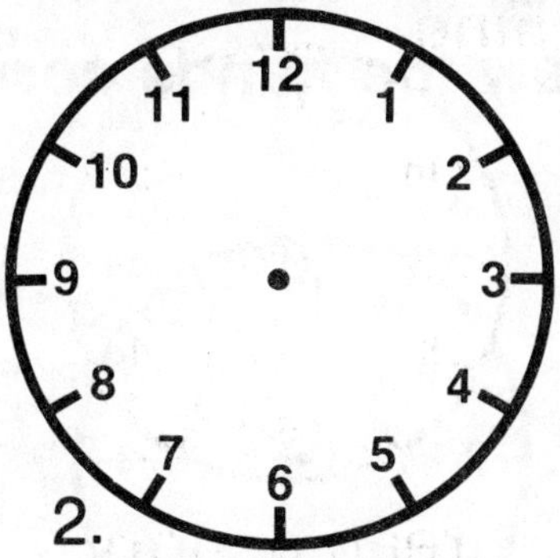
2. eat dinner 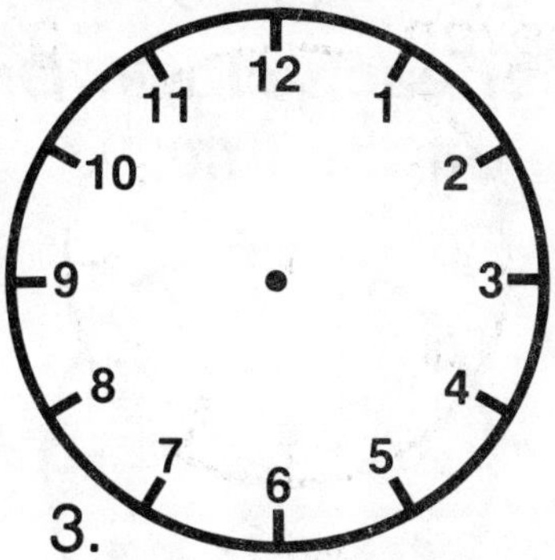
3. play outside 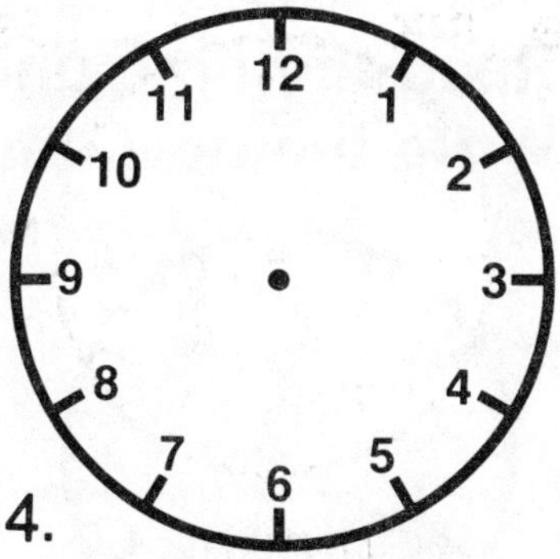
4. get out of bed

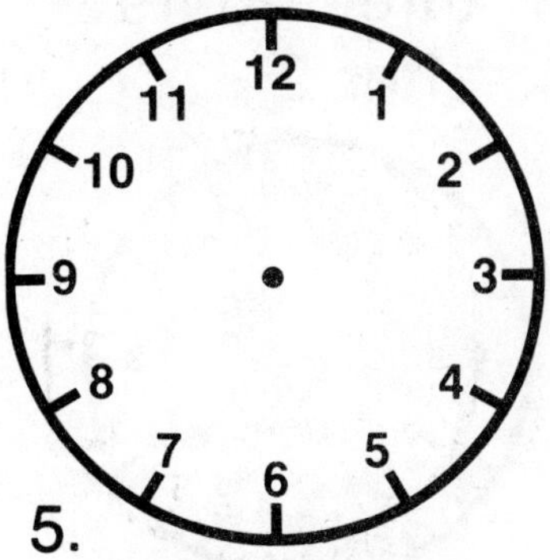

5. go to school
6. watch TV 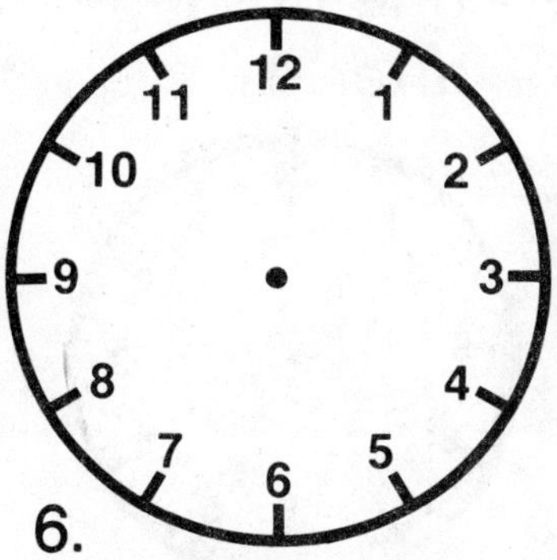
7. eat breakfast 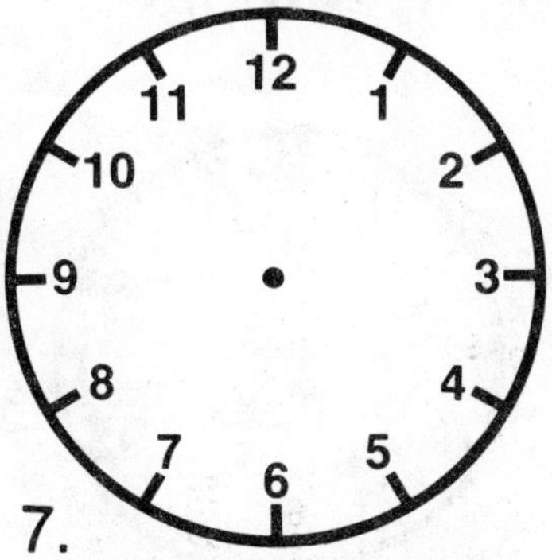
8. read a story 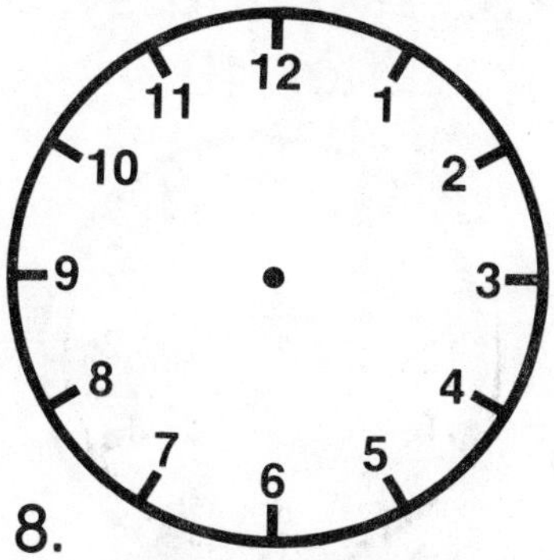

© Frank Schaffer Publications, Inc.   3   FS-2674 MATH WORKBOOK-Time and Money

Name _____ Skill: Hour

Hour hand   Minute hand

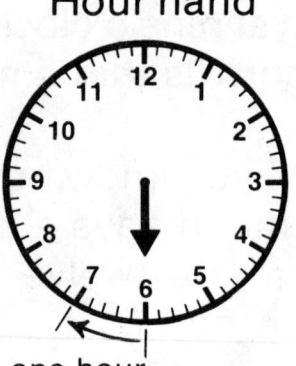
one hour

There are 60 minutes in one hour.

The hour is three o'clock or 3:00.

Fill in the missing hands.

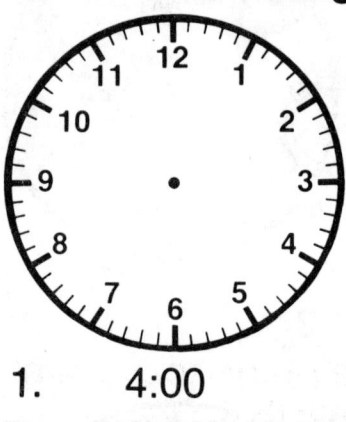

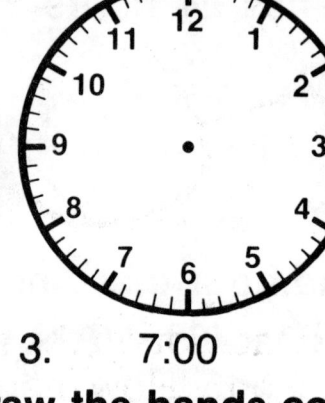

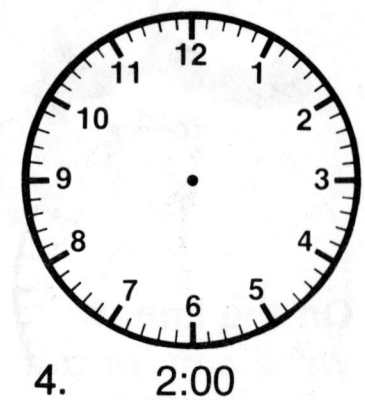

1. 4:00   2. 11:00   3. 7:00   4. 2:00

Read the time under each clock. Draw the hands correctly.
Write the new time on the line.

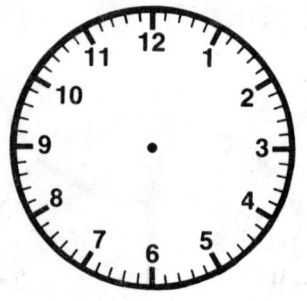      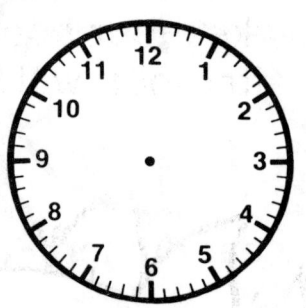   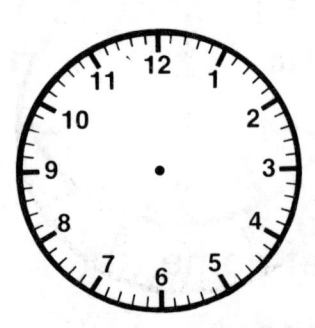

5. two hours after 10   6. five hours after 6   7. one hour after 9   8. three hours after 2

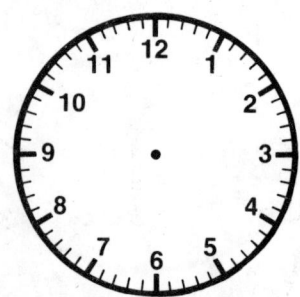      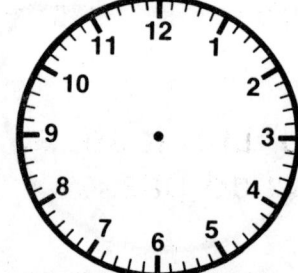   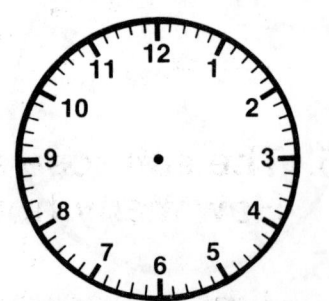

9. one hour before 11   10. four hours before 4   11. six hours before 10   12. two hours before 2

© Frank Schaffer Publications, Inc.   FS-2674 Math Workbook-Time and Money

Name _____   Skill: Hour

Read the sentences and write your answers. The time is on the hour.

1. Sue gets up at 7 a.m. In one hour, she will eat breakfast. The time will be _____ .

2. Mr. Loof is coming at 2 p.m. It will take him three hours to fix the TV. He will finish at _____ .

3. Dad got on the plane at 12 p.m. The plane took off at 2 p.m. How many hours did Dad have to wait? _____

4. We are driving to Magic Land. We'll leave at 9 a.m. It takes four hours to get there. What time in the afternoon will we arrive? _____

5. The sun rose at 6 a.m. It set at 5 p.m. How many hours had passed? _____

6. Jimmy's game started at 6 p.m. It ended at 9 p.m. How long did the game last? _____

© Frank Schaffer Publications, Inc.   5   FS-2674 Math Workbook-Time and Money

Name _____ Skill: Half hour

What time is it?

1. _____ 2. _____ 3. _____ 4. _____

5. _____ 6. _____ 7. _____ 8. _____

9. _____ 10. _____ 11. _____ 12. _____

13. _____ 14. _____

© Frank Schaffer Publications, Inc. FS-2674 Math Workbook-Time and Money

Name _____   Skill: Half hour

Half hour = 30 minutes

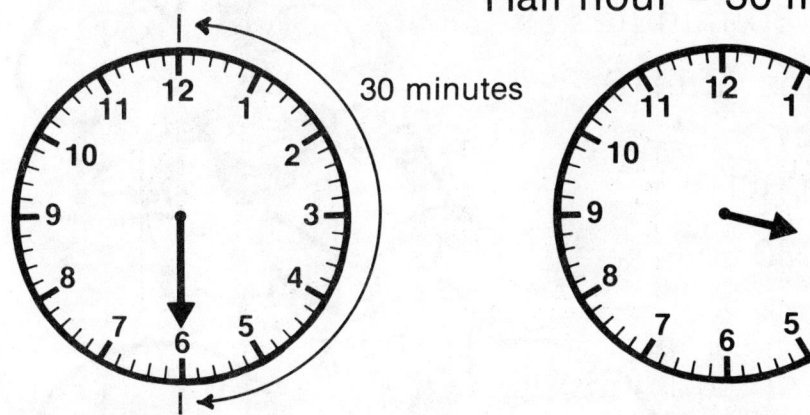

The time is 30 minutes past . . . the hour of 3 o'clock.   The time is 3:30 **or** half past 3.

Draw the hands to show the time.

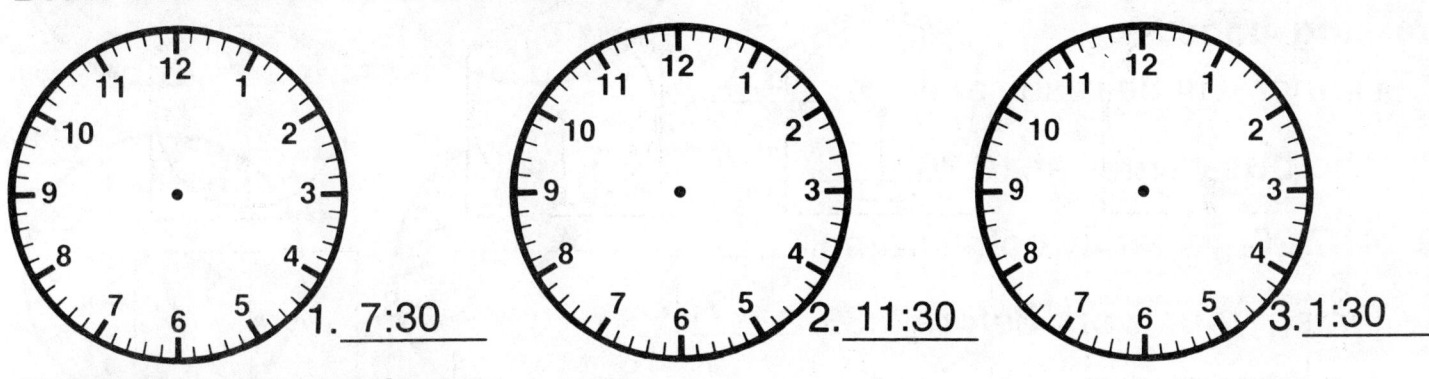

1. 7:30   2. 11:30   3. 1:30

Write the correct time.

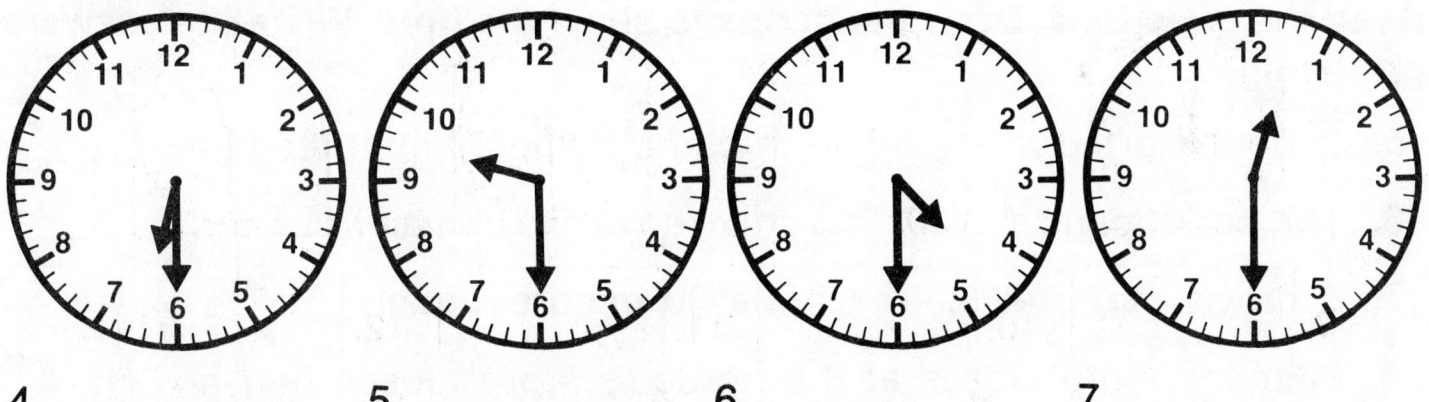

4. _____   5. _____   6. _____   7. _____

Answer the questions.

8. Jan ate lunch at 12 p.m. Half an hour later she went out to play. What time is it? _____

9. "Huckle Harry" ends at 9 p.m. John goes to bed 30 minutes later. What time will it be? _____

10. Mom leaves for work at 6 a.m. She gets to work half an hour later. What time is it? _____

Name _____   Skill: Half hour

Half hour = 30 minutes
30 minutes after 9
or
half past 9
9:30

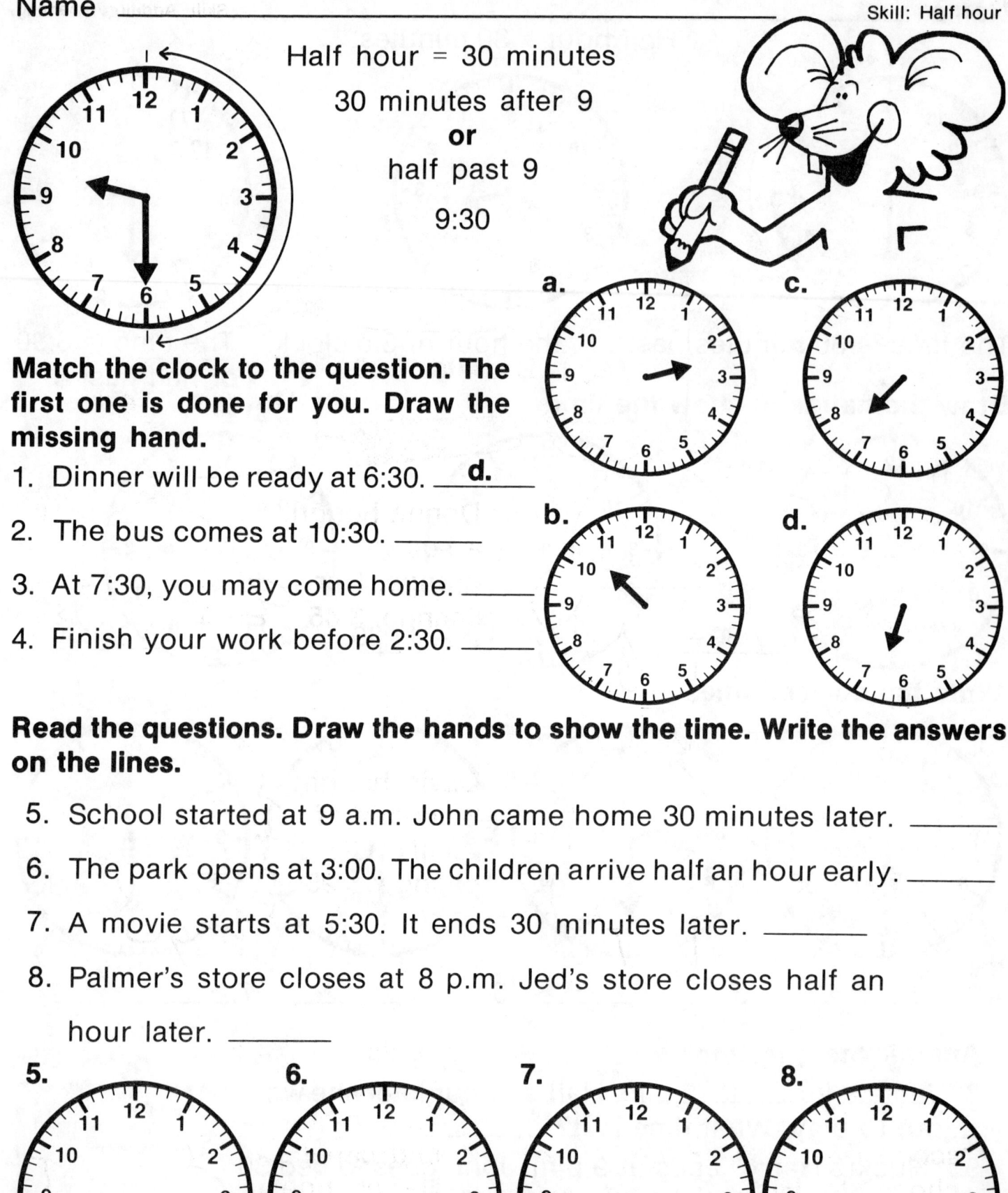

Match the clock to the question. The first one is done for you. Draw the missing hand.

1. Dinner will be ready at 6:30. __d.__
2. The bus comes at 10:30. _____
3. At 7:30, you may come home. _____
4. Finish your work before 2:30. _____

Read the questions. Draw the hands to show the time. Write the answers on the lines.

5. School started at 9 a.m. John came home 30 minutes later. _____
6. The park opens at 3:00. The children arrive half an hour early. _____
7. A movie starts at 5:30. It ends 30 minutes later. _____
8. Palmer's store closes at 8 p.m. Jed's store closes half an hour later. _____

Name _____   Review: Hour and half hour

What time is it?

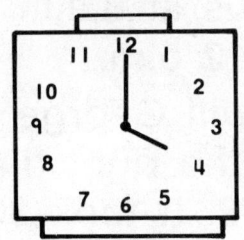 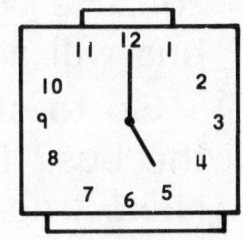 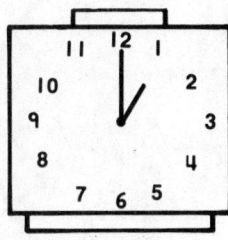

1. _____   2. _____   3. _____   4. _____

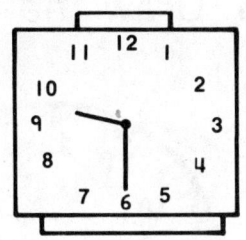 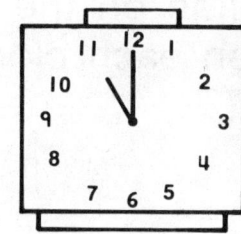 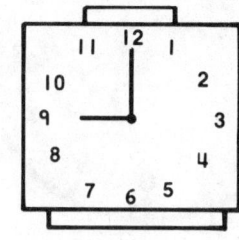 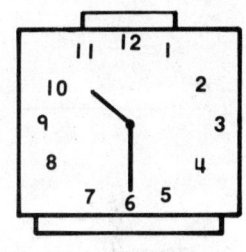

5. _____   6. _____   7. _____   8. _____

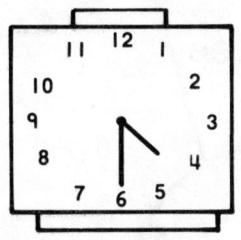 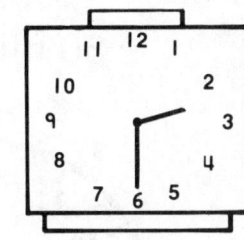 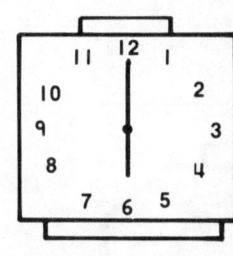 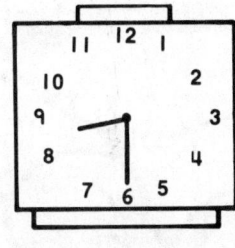

9. _____   10. _____   11. _____   12. _____

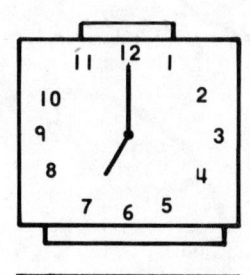 

13. _____   14. _____

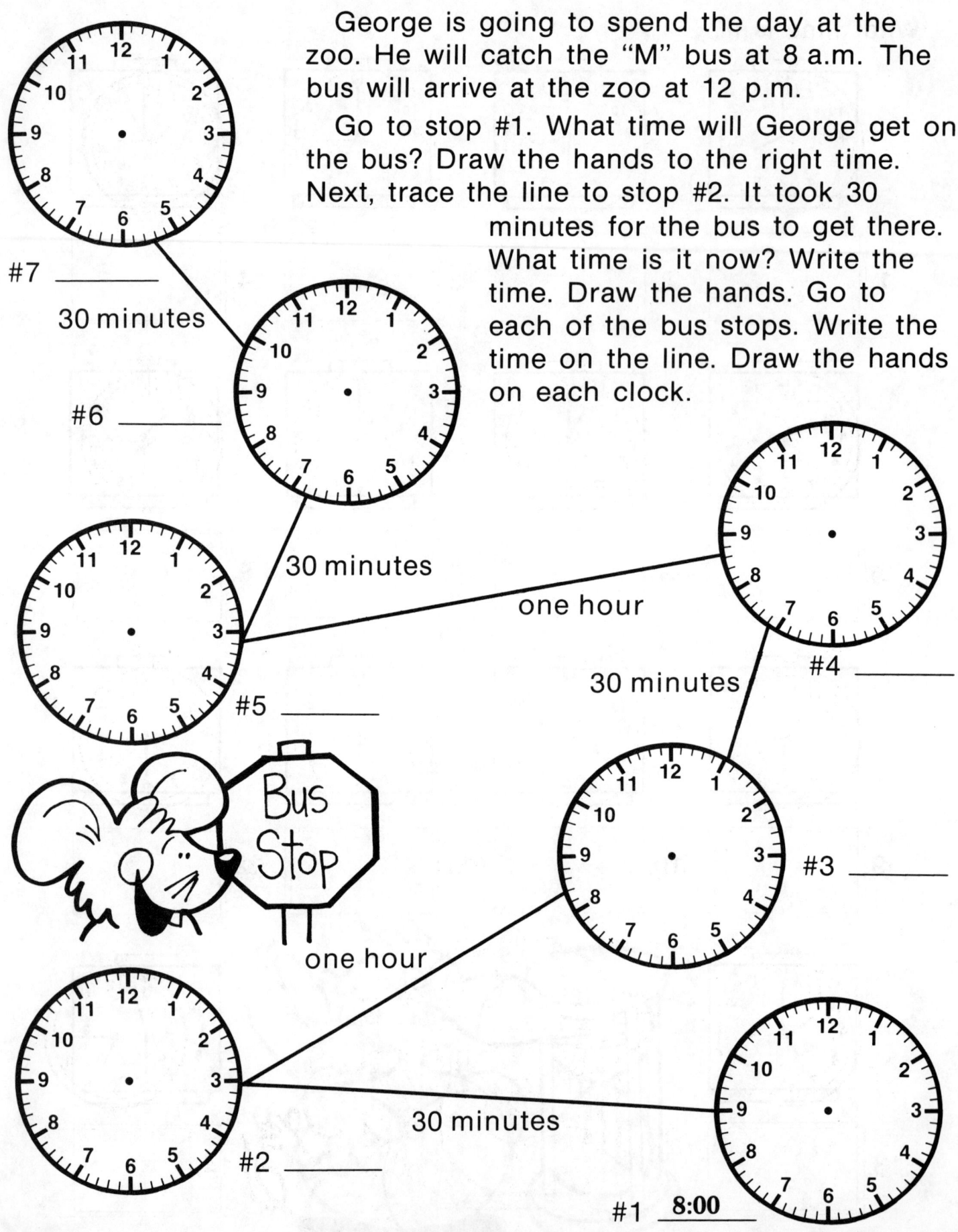

Name _____   Skill: Fifteen past the hour

What time is it?

1. _____   2. _____   3. _____   4. _____

5. _____   6. _____   7. _____   8. _____

9. _____   10. _____   11. _____   12. _____

13. _____   14. _____

Name _____   Skill: Fifteen past the hour

1.  2.  3.

The time is fifteen **minutes** past . . . the **hour** of four.   The time is 4:15 **OR** quarter past four

Draw the hands to the correct time, then write the time as in clock number 3.

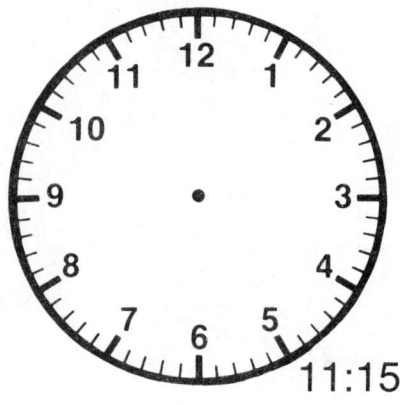 11:15   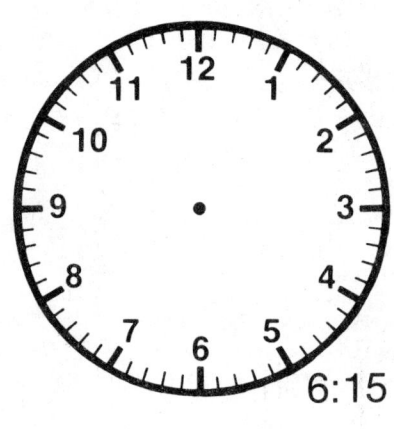 6:15   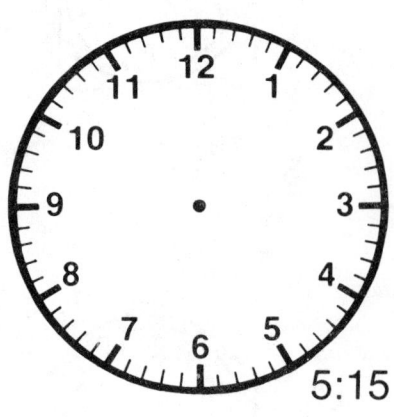 5:15

1. _____   2. _____   3. _____

Answer the questions.

4. Jane ate lunch at noon. Fifteen minutes later she took a nap. The time is _____ .

5. The play started at 5:30. Everyone came fifteen minutes early. What time is it? _____

6. I'm taking the bus to the dentist. It leaves at 9:15. I'll arrive two hours later. The time will be _____ .

7. Mike got to the game at 6:00. The game started fifteen minutes later. What time did the game start? _____

Name _____   Review

What time is it?

1. ____ 2. ____ 3. ____ 4. ____

   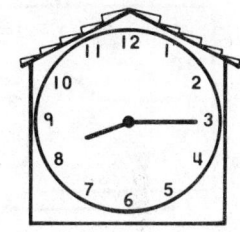

5. ____ 6. ____ 7. ____ 8. ____

9. ____ 10. ____ 11. ____ 12. ____

13. ____ 14. ____

Name _____   Skill: Fifteen before the hour

What time is it?

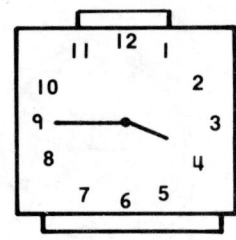 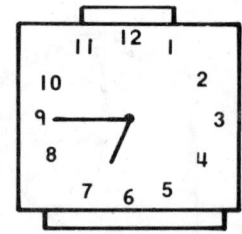 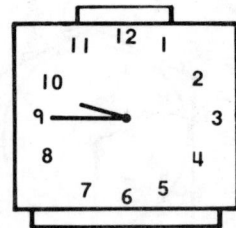

1. 2. 3. 4.

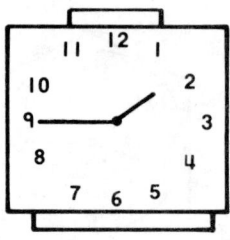 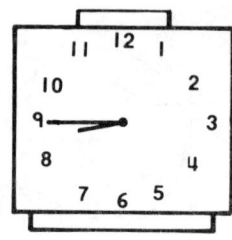 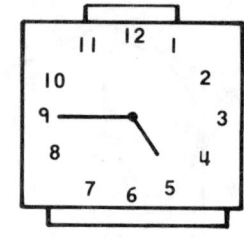 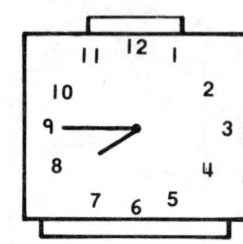

5. 6. 7. 8.

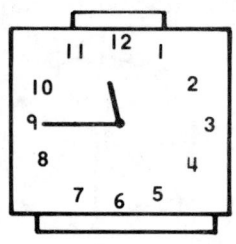 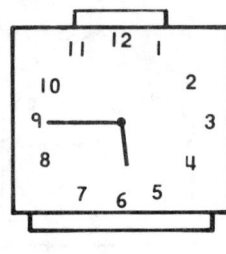 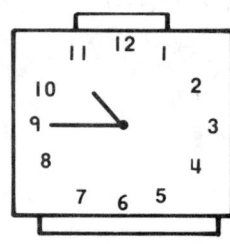 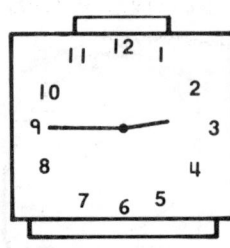

9. 10. 11. 12.

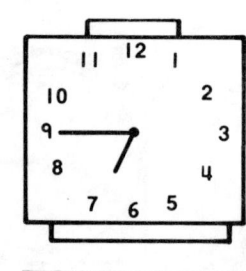  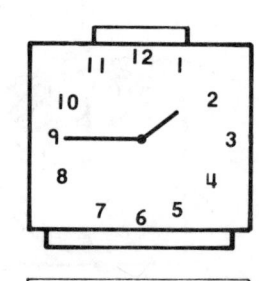

13. 14.

© Frank Schaffer Publications, Inc.   FS-2674 Math Workbook-Time and Money

Name _____ Skill: Fifteen before the hour
(mins. = minutes)

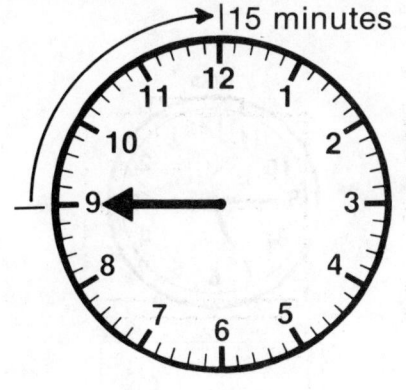  

The time is fifteen
minutes **before** . . . the hour of seven   OR quarter to seven

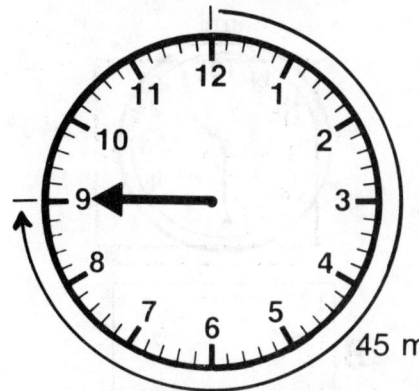  

The time is 45 minutes **after** . . . the hour of six OR 6:45

Look at the time under each clock. Draw the hands in the right places.

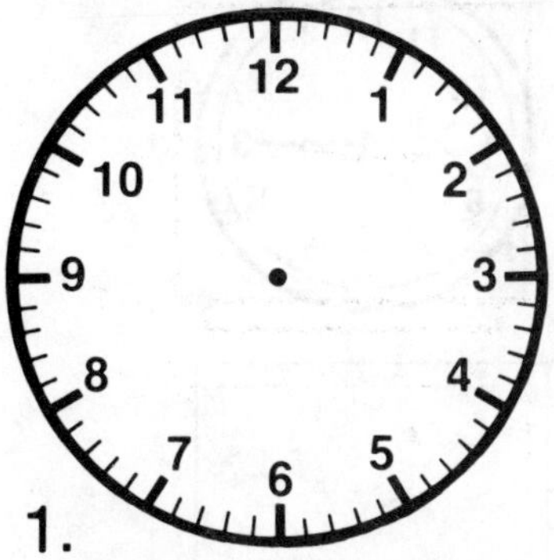

1. quarter to one 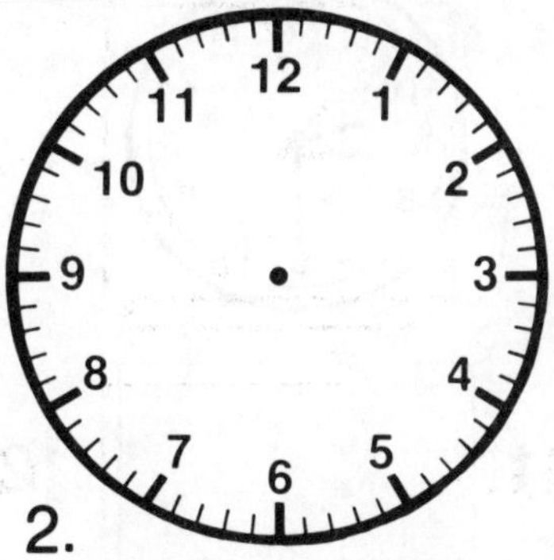
2. 45 minutes after four 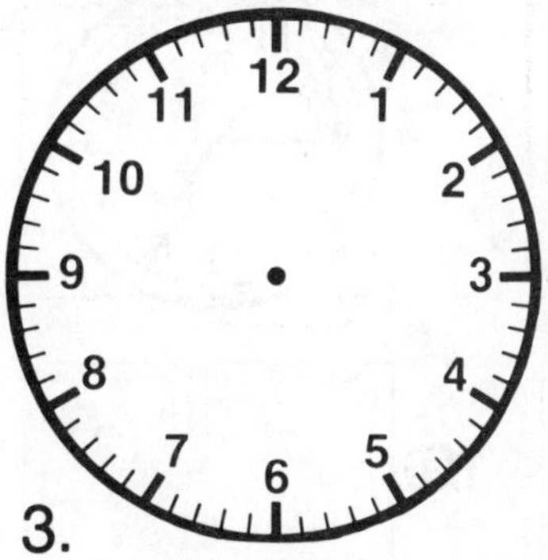
3. 10:45 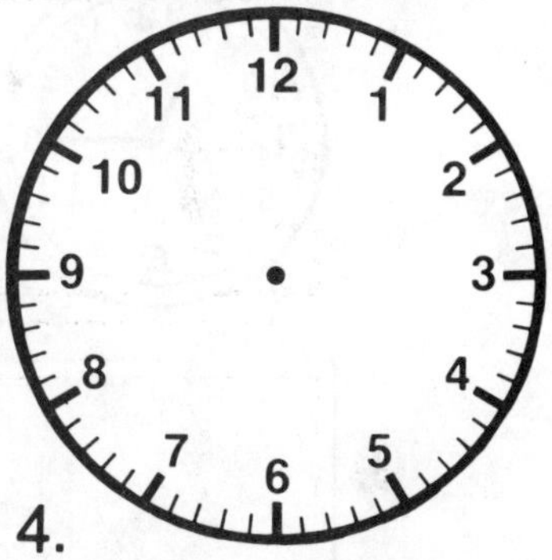
4. quarter to six

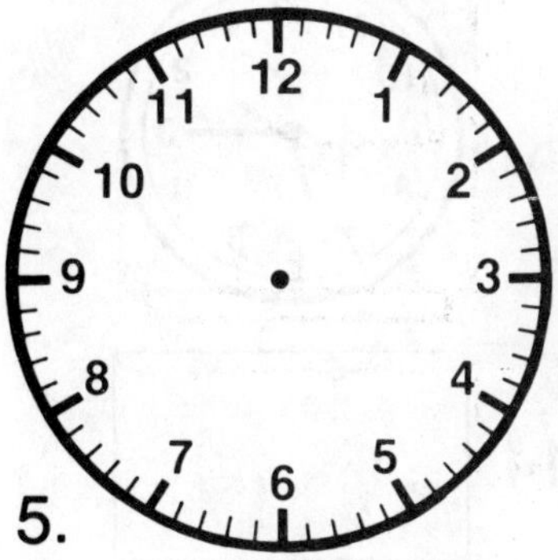

5. 2:45
6. 8:45 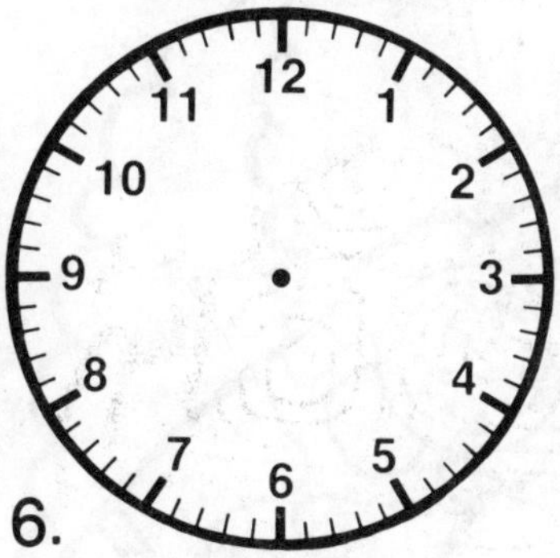
7. 45 mins. after two 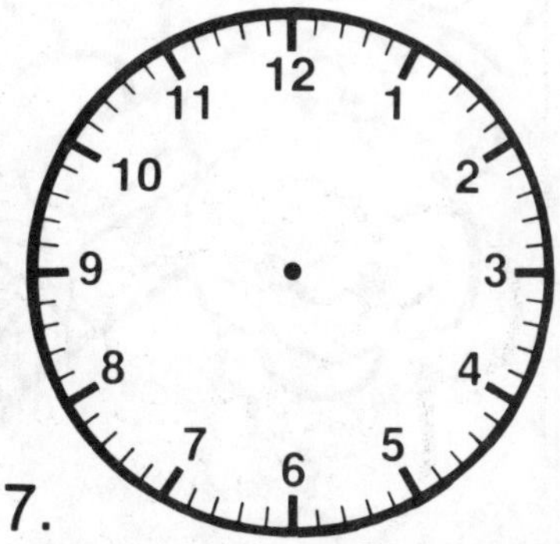
8. quarter to five 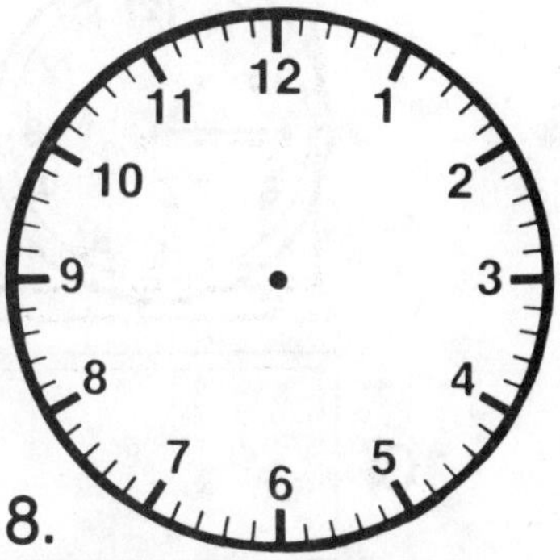

© Frank Schaffer Publications, Inc.   15   FS-2674 Math Workbook-Time and Money

Name _____   Skill: Review

What time is it?

1.    2. 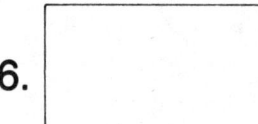   3. 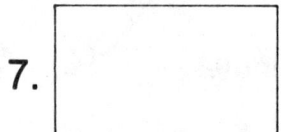   4.

5.   6.   7.   8.

9. 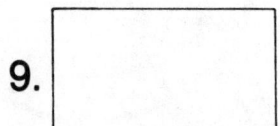   10. 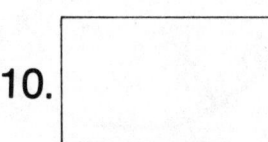   11. 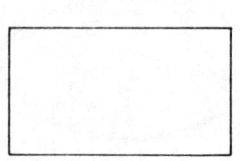   12.

13.    14.

© Frank Schaffer Publications, Inc.   FS-2674 Math Workbook-Time and Money

Name _____ Skill: Increments of ten

What time is it?

1. 2. 3. 4.

5. 6. 7. 8.

9. 10. 11. 12.

13. 14.

© Frank Schaffer Publications, Inc.   17   FS-2674 Math Workbook-Time and Money

Name _____

Skill: Increments of ten

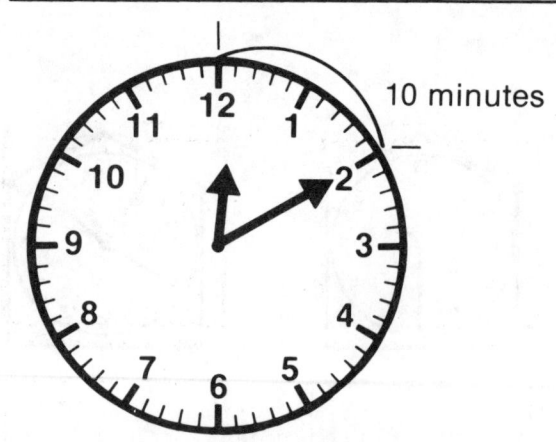

The time is ten minutes past the hour of 12.

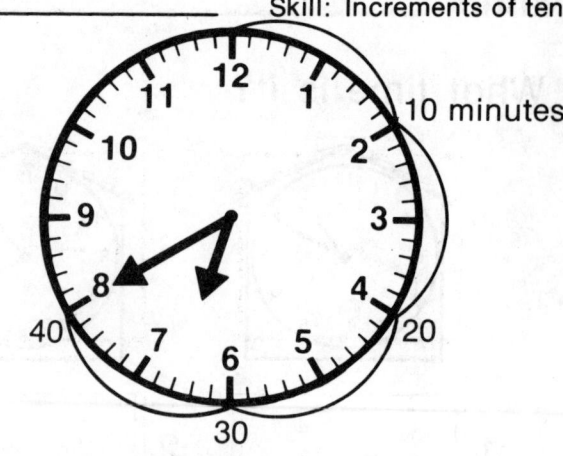

The time is forty minutes past the hour of 6.

Look at the time under each clock. Draw the hands in the right places.

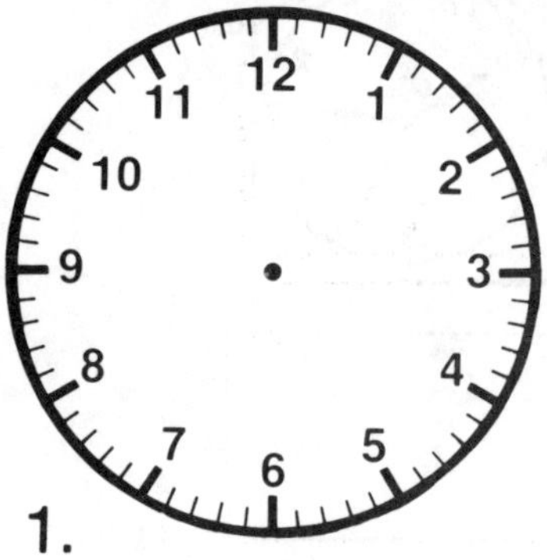

1. 9:40 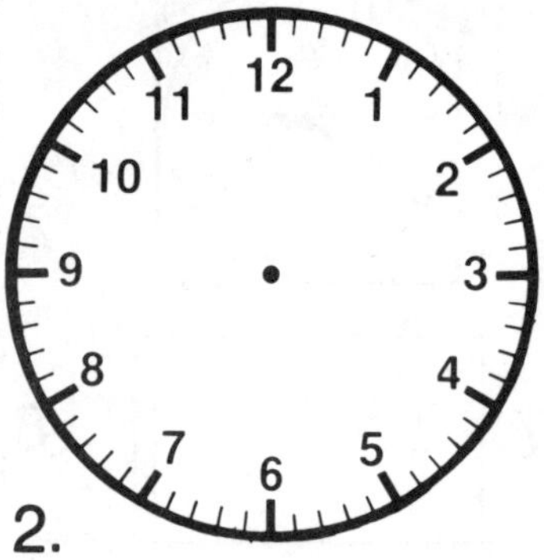
2. fifty minutes after 8 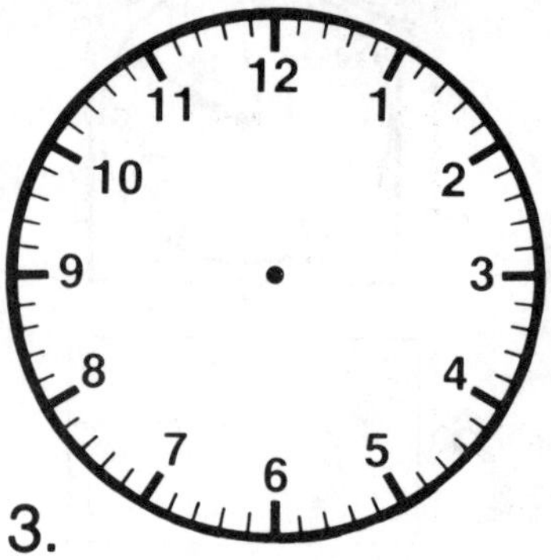
3. 11:20 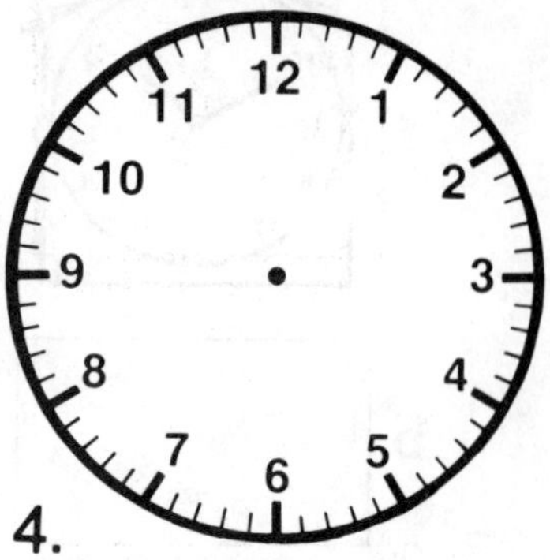
4. ten minutes after 4

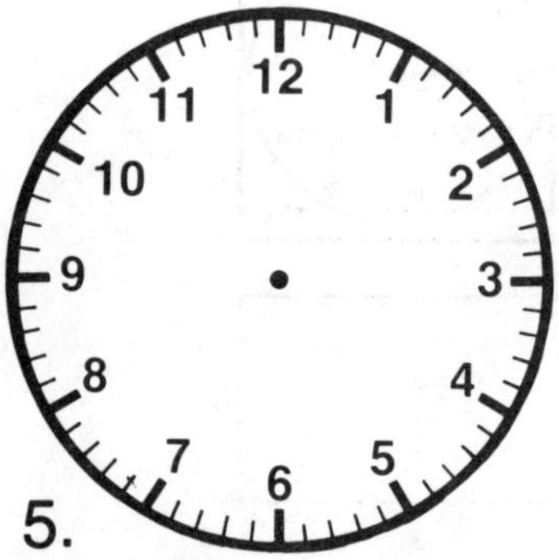

5. twenty minutes after 4
6. 10:50 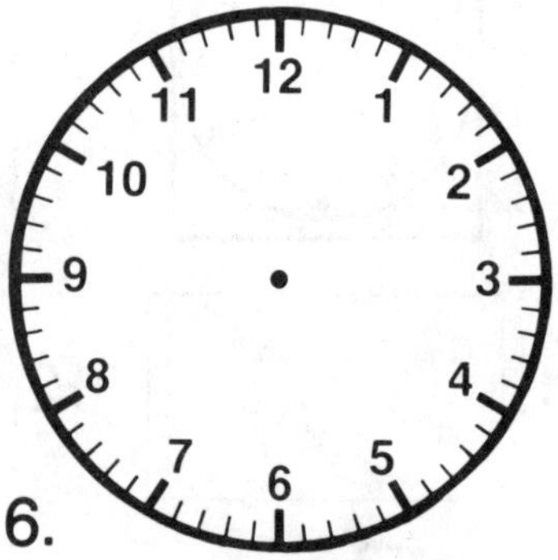
7. forty minutes after 6 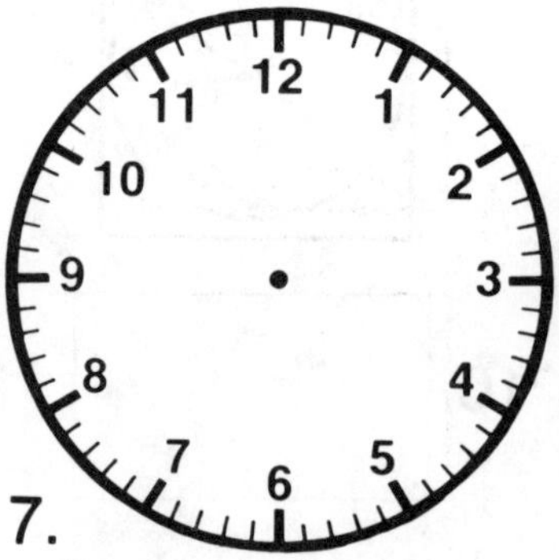
8. 12:30 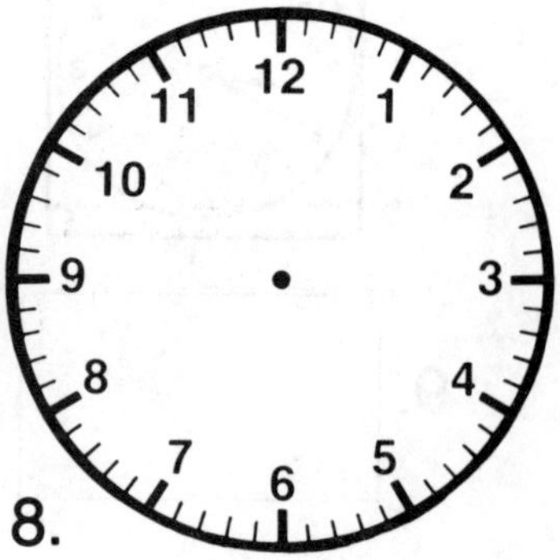

Count by ten. Write how many minutes there are:

9. from 4:20 to 4:50 _____

10. from 11:40 to 12:20 _____

11. from 7:50 to 8:10 _____

Name _____ Skill: Increments of five

What time is it?

 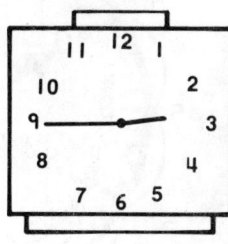 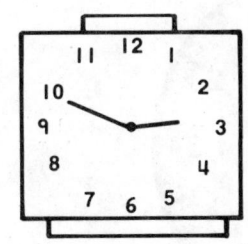 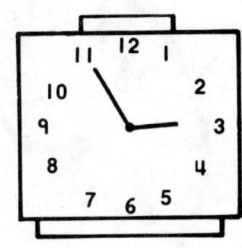

1. _____ 2. _____ 3. _____ 4. _____

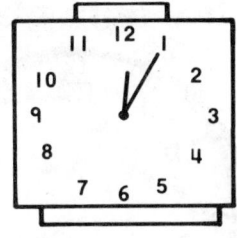 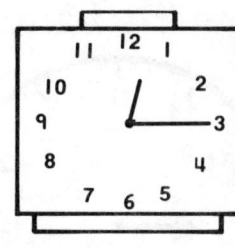 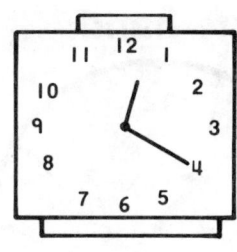 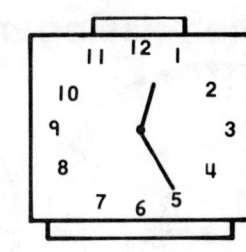

5. _____ 6. _____ 7. _____ 8. _____

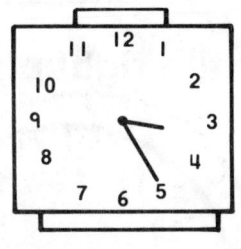 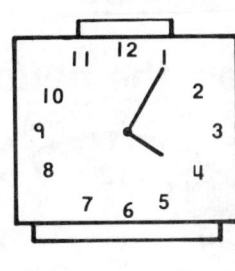 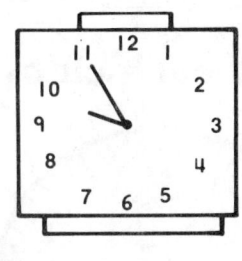 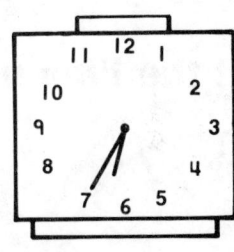

9. _____ 10. _____ 11. _____ 12. _____

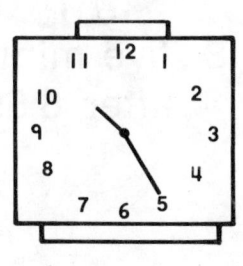 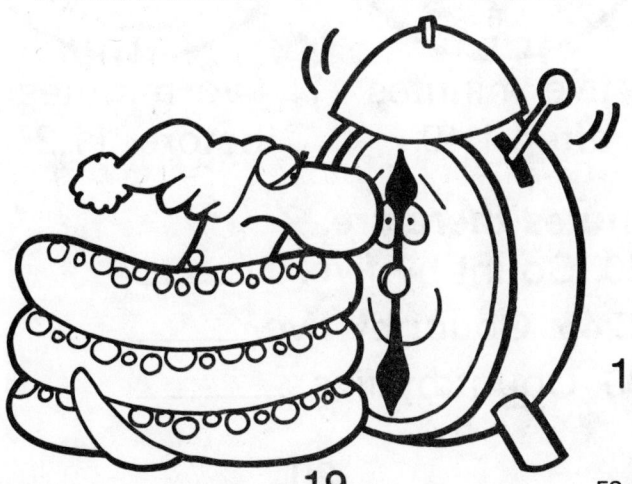

13. _____ 14. _____

© Frank Schaffer Publications, Inc. FS-2674 Math Workbook-Time and Money

Name _____   Skill: Increments of five

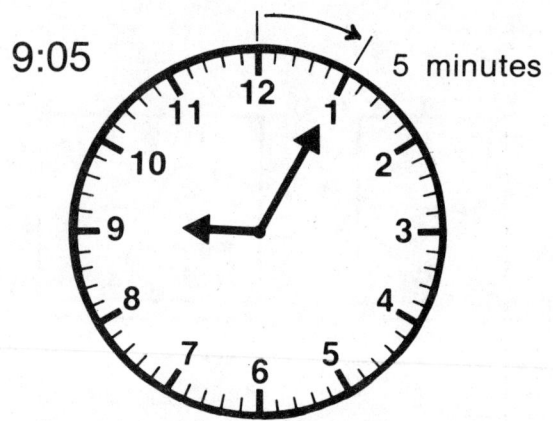
9:05

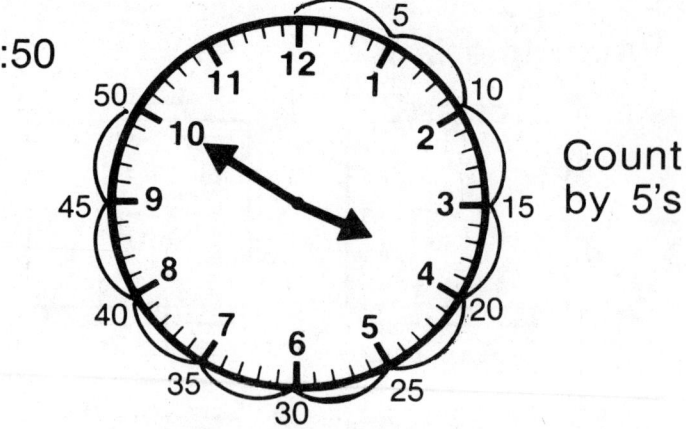
3:50 Count by 5's

The time is five minutes past the hour of 9.

The time is fifty minutes past the hour of 3.

Draw the missing minute hand.

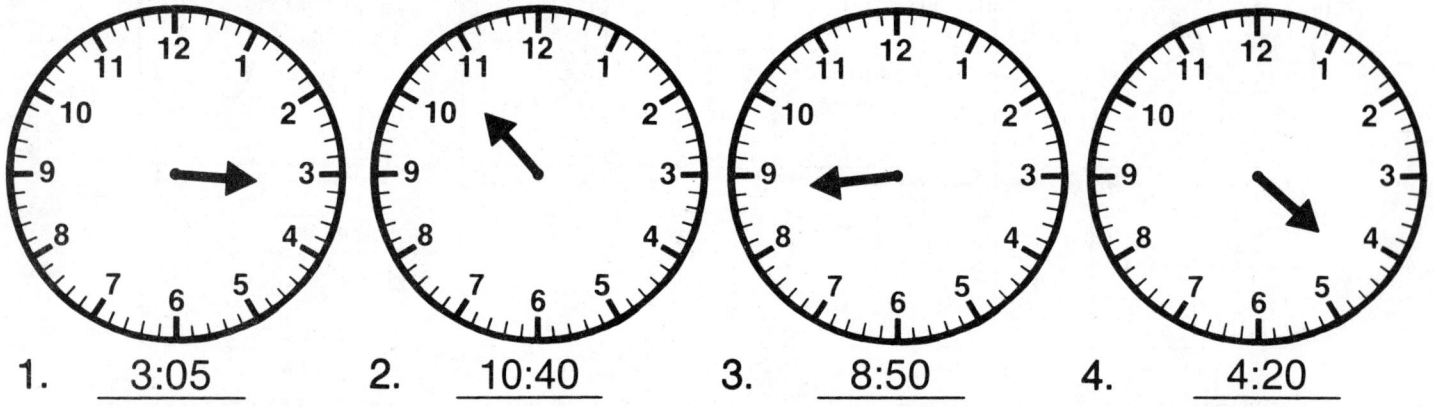

1. 3:05   2. 10:40   3. 8:50   4. 4:20

Look at the time under each clock. Draw the hands in the right places.

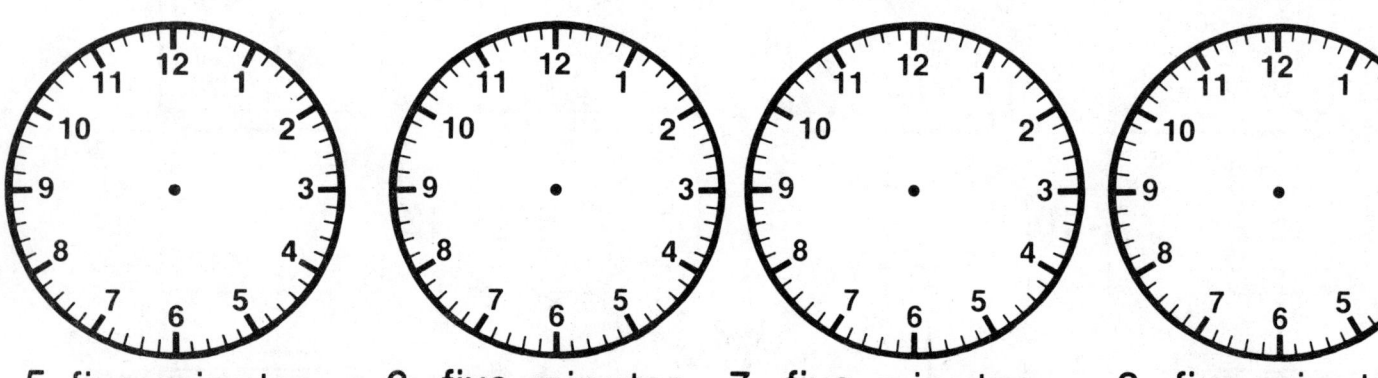

5. five minutes before 7:45
6. five minutes after 1:10
7. five minutes before 11:25
8. five minutes after 6:55

Write how many minutes there are:
 9. from 9:30 to 9:50. Count by five. _____
10. from 12:15 to 12:45. Count by five. _____
11. from 2:35 to 3:00. Count by five. _____

Name _____ Skill: Review

Time for:

lunch

school

bed

1. 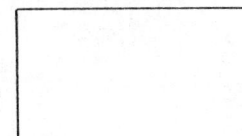 2. 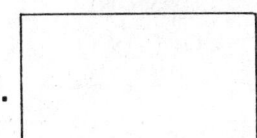 3. 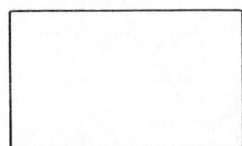 4.

dinner

play

a party

reading

math

5. 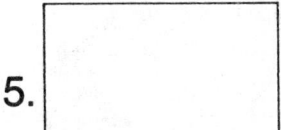 6.  7. 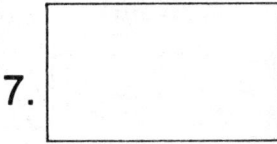 8.

a game

breakfast

TV

waking up

9. 10. 11. 12.

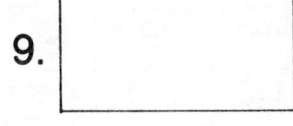
singing

milk and cookies

13.  14.

Name _____ Review

What time is it?

ten past five | five to six | fifteen past two | five past eight

1. ____ 2. ____ 3. ____ 4. ____

six-thirty | twenty past nine | fifteen to three | two thirty-five

5. ____ 6. ____ 7. ____ 8. ____

quarter to seven | ten to eleven | twenty to four | fifteen to two

9. ____ 10. ____ 11. ____ 12. ____

five past seven | | twenty-five past twelve

13. ____ 14. ____

© Frank Schaffer Publications, Inc.   FS-2674 Math Workbook—Time and Money

Name _____   Review

What time is it?

 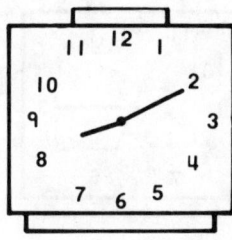  

1. ____ 2. ____ 3. ____ 4. ____

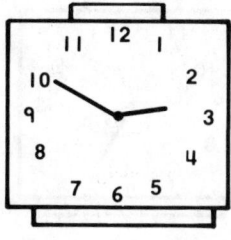   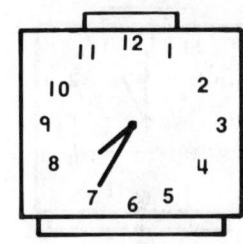

5. ____ 6. ____ 7. ____ 8. ____

  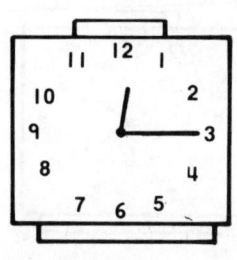 

9. ____ 10. ____ 11. ____ 12. ____

13. ____ 14. ____

Name _____   Review

What time is it?

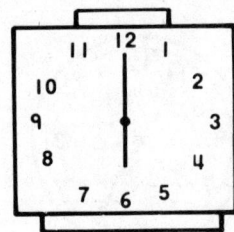  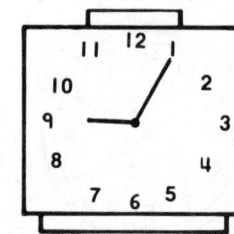 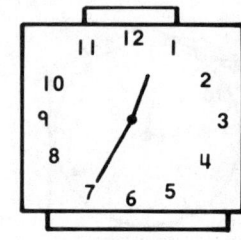 

1. _____   2. _____   3. _____   4. _____

 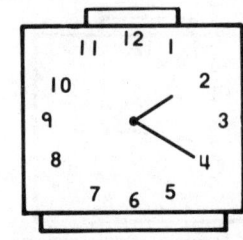 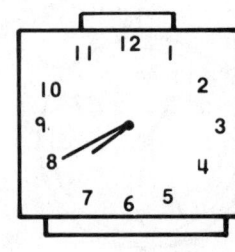 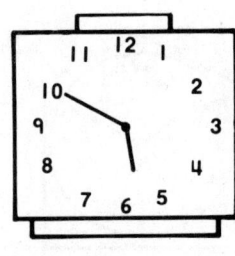

5. _____   6. _____   7. _____   8. _____

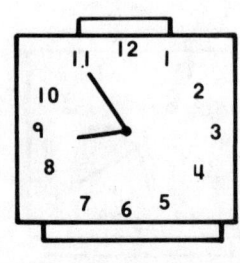 

9. _____   10. _____   11. _____   12. _____

13. _____   14. _____

Name _____ Review

What time is it?

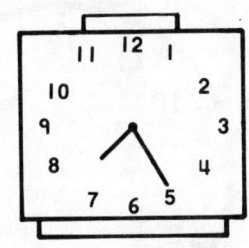   

1. _____ 2. _____ 3. _____ 4. _____

  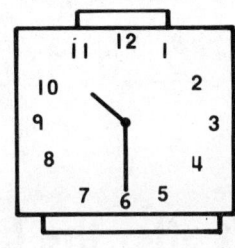 

5. _____ 6. _____ 7. _____ 8. _____

 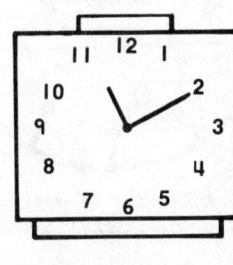  

9. _____ 10. _____ 11. _____ 12. _____

13. _____ 14. _____

Name _____   Review

Look at the time under each clock. Draw the hands in the right places.

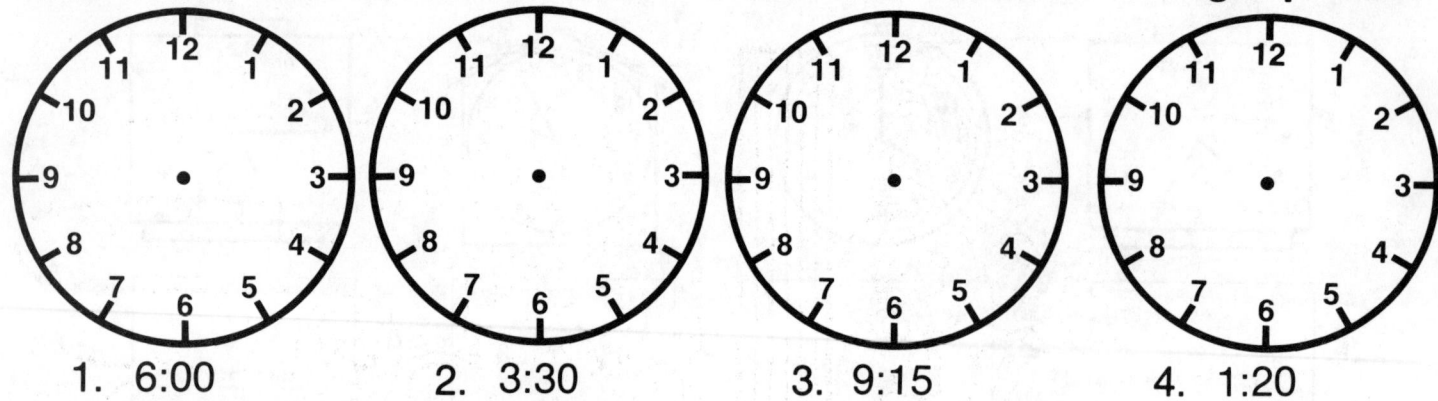

1. 6:00   2. 3:30   3. 9:15   4. 1:20

Write these times using numbers.

5. seven forty-five _____   7. quarter after two _____

6. half past four _____   8. six thirty _____

How many <u>minutes</u> is it after the hour?

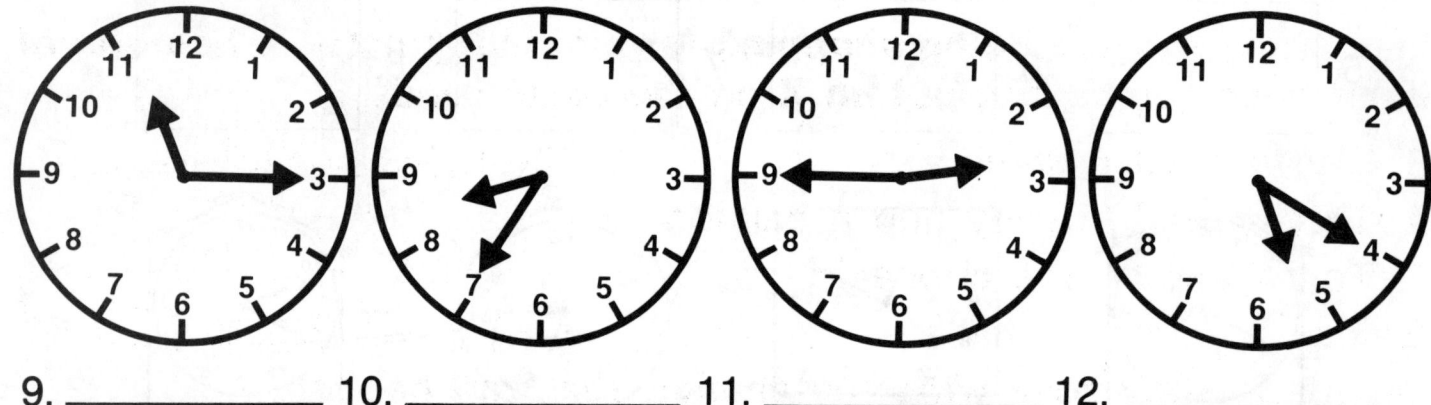

9. _____   10. _____   11. _____   12. _____

Look at the clocks. What time will it be:

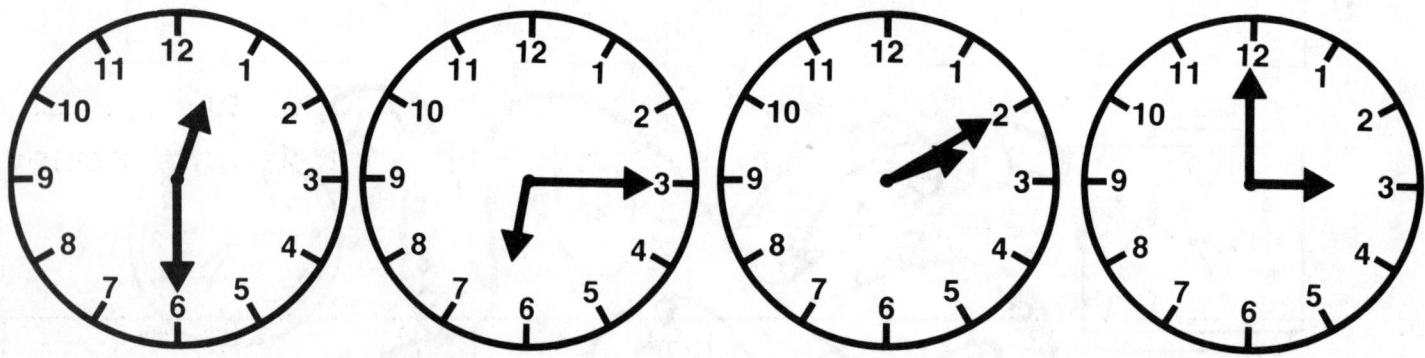

two hours later   four hours before   one hour later   seven hours before

13. _____   14. _____   15. _____   16. _____

Name _____ Skill: Dimes and nickels

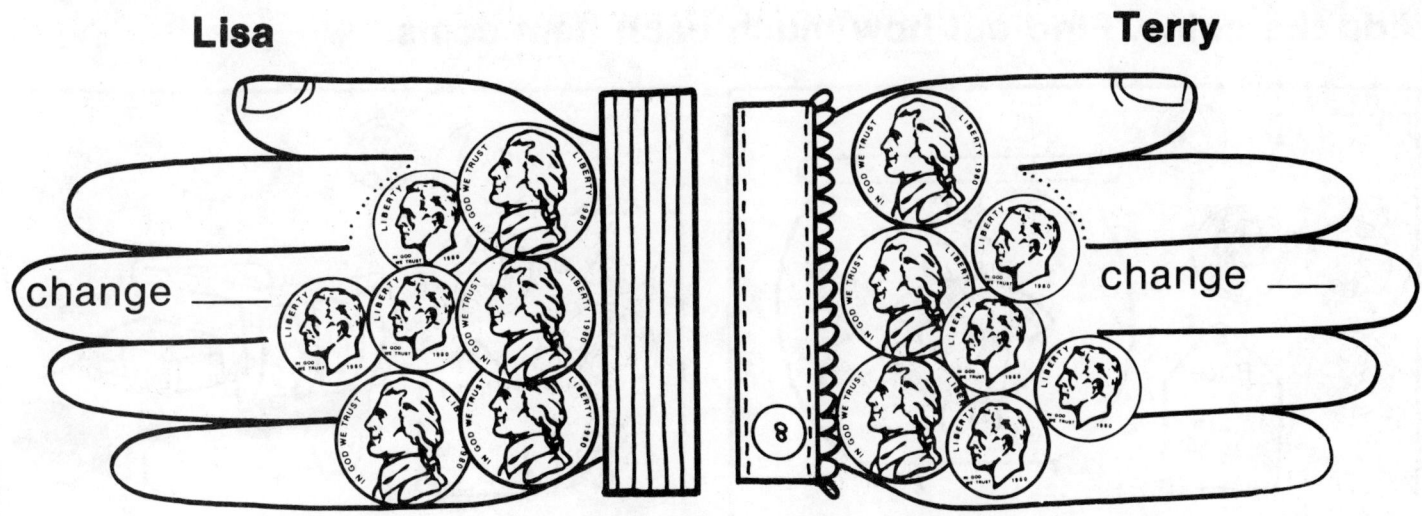

How much money does each person have?

| 1. ___ dimes = ___ ¢
 ___ nickels = ___ ¢ | Total ___ ¢ | 2. ___ dimes = ___ ¢
 ___ nickels = ___ ¢ | Total ___ ¢ |

Lisa and Terry are going shopping for a birthday party. When one of them buys something, put an X on the coins used.

3. Lisa bought a game.
4. Terry bought candy and a balloon.
5. Terry bought one cupcake.
6. Lisa bought one hat.

On the hands above write the change each girl has left.

7. Terry wants to buy a candle. How much more money does she need? _____ ¢
8. How much money do Lisa and Terry now have together? _____ ¢
9. The girls want to buy ice cream. It costs 25¢. How much more money will they need to buy it? _____ ¢

Name _____ Skill: Dimes, nickels, pennies

Add the coins. Find out how much each item costs.

Find the number of dimes and nickels there are in the starred (*) numbers. Add the cost. Write the price on the tag.

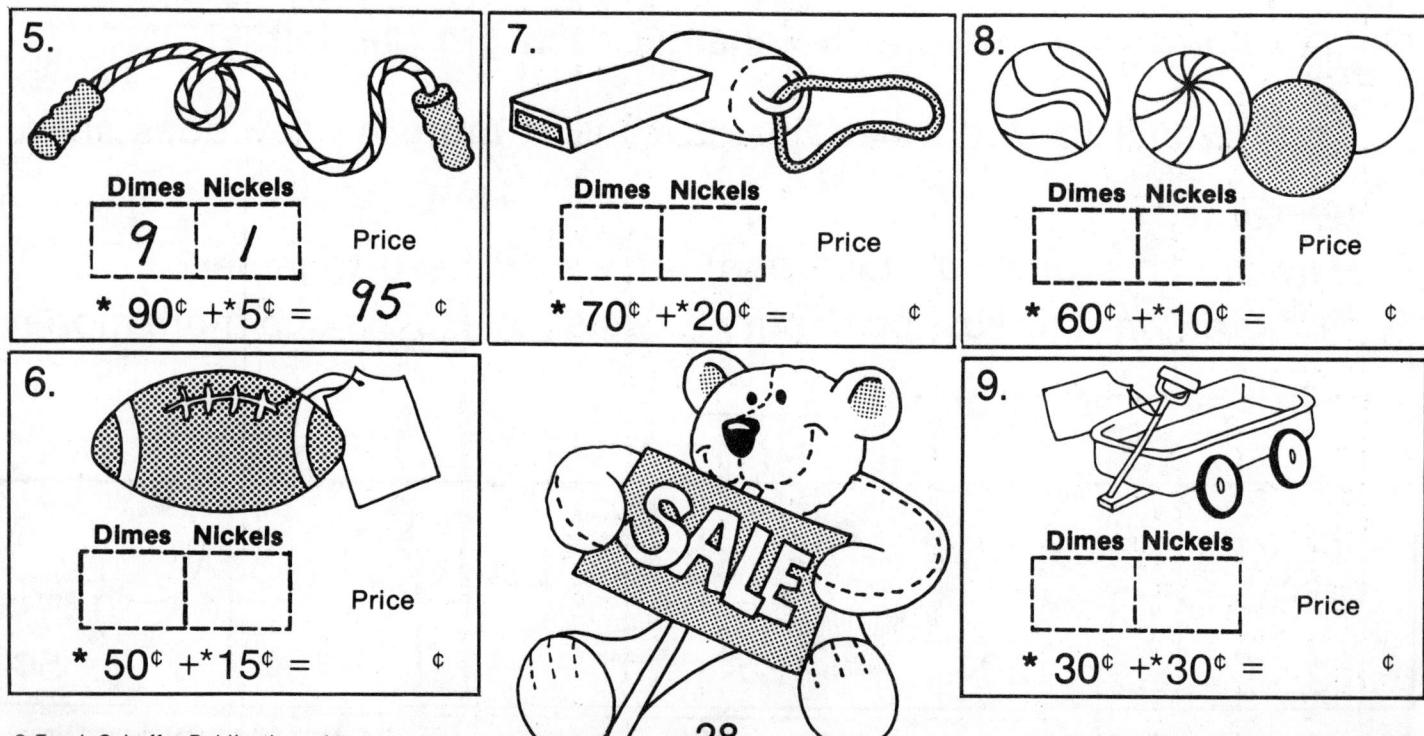

© Frank Schaffer Publications, Inc.   FS-2674 Math Workbook-Time and Money

Name _____ Skill: Counting by 5's and 1's

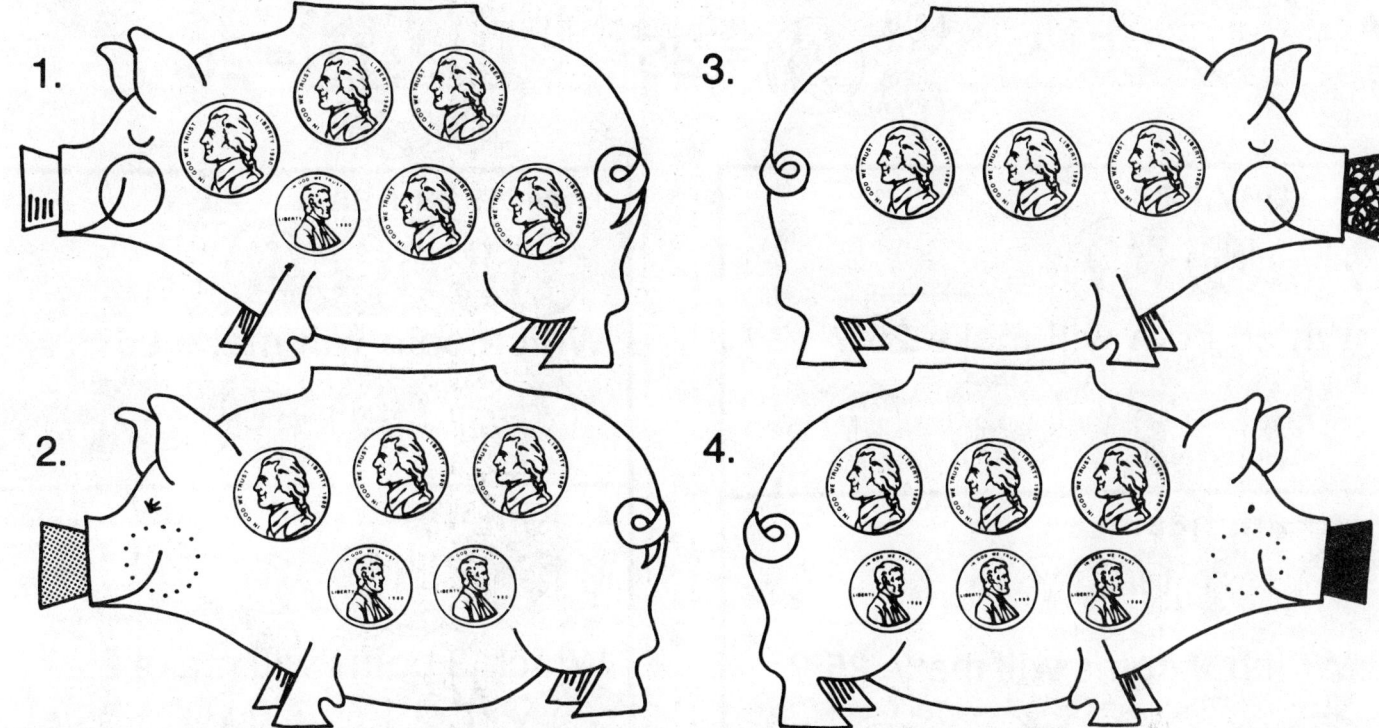

Add the coins to find out how much is in each bank.   Total

1. 5 +	=	¢
2.	=	¢
3.	=	¢
4.	=	¢

How many:

5. (nickel) in 25¢? ____
6. (penny) in 3¢? ____
7. (nickel) in 30¢? ____
8. (penny) in 9¢? ____

9. (nickel) in 10¢? ____
10. (penny) in 8¢? ____
11. (nickel) in 40¢? ____
12. (penny) in 1¢? ____

13. (nickel) in 50¢? ____
14. (penny) in 6¢? ____
15. (nickel) in 35¢? ____
16. (penny) in 4¢? ____

Name _____   Skill: Quarters

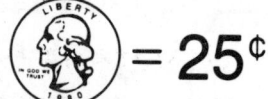  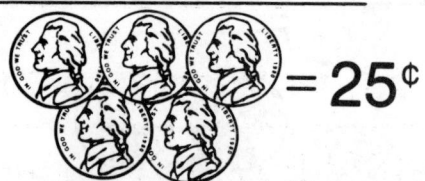

1. Tony has:
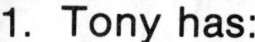 = _____ ¢

Which coin will make 25¢?

2. Paul has:
 = _____ ¢

Which 2 coins will make 25¢?

3. Monica has:
 = _____ ¢

Which coin will make 25¢?

4. Kim has:
_____ ¢

Which 3 coins will make 25¢?

5. Monica has ⓞⓞⓞ. She bought a 🪆20¢. How much does she have left? _____ ¢

6. Paul has ⓟⓟⓟⓟⓟⓝ. He gave Kim 15¢. How much money does Paul have now? _____ ¢

7. Tony has ⓝⓝⓝⓝⓝ. He put 25¢ in the bank. How much is left? _____ ¢

8. Kim wants 🛼25¢. She has ⓞⓞ. How much more money does she need? _____ ¢

9. Paul has ⓠ. He wants 🫧5¢ and a 🎺15¢. Can he buy both? _____ How much change will he have? _____ ¢

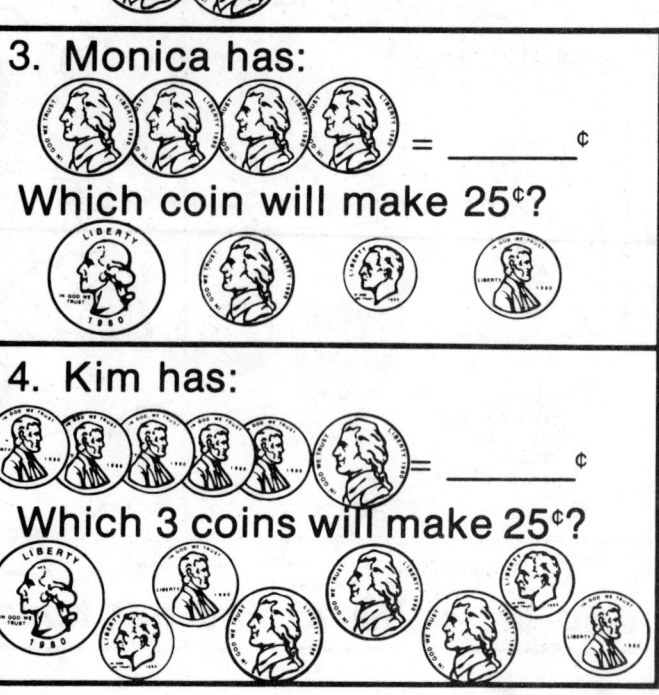

© Frank Schaffer Publications, Inc.   30   FS-2674 MATH WORKBOOK-Time and Money

Name _____ Skill: Adding quarters, dimes, nickels

The Bailey family went out to dinner at Don's Deli. They all chose what they wanted in their sandwiches. Add the cost of each sandwich. Follow the example.

Deli Case

sausage-25¢   lettuce-10¢   pickles-10¢   olives-5¢   meatballs-25¢   salami-10¢   cheese-5¢

Example:

1. __25¢__ + __10¢__ + __5¢__ + __5¢__ = 45¢

2. _____ + _____ + _____ + _____ + _____ =

3. _____ + _____ + _____ + _____ + _____ =

4. _____ + _____ + _____ + _____ + _____ =

Name _____   Skill: Half dollar

1. Count by 5's   2. Count by 10's   3. Count by 25's

= _____ ¢   = _____ ¢   = _____ ¢

4.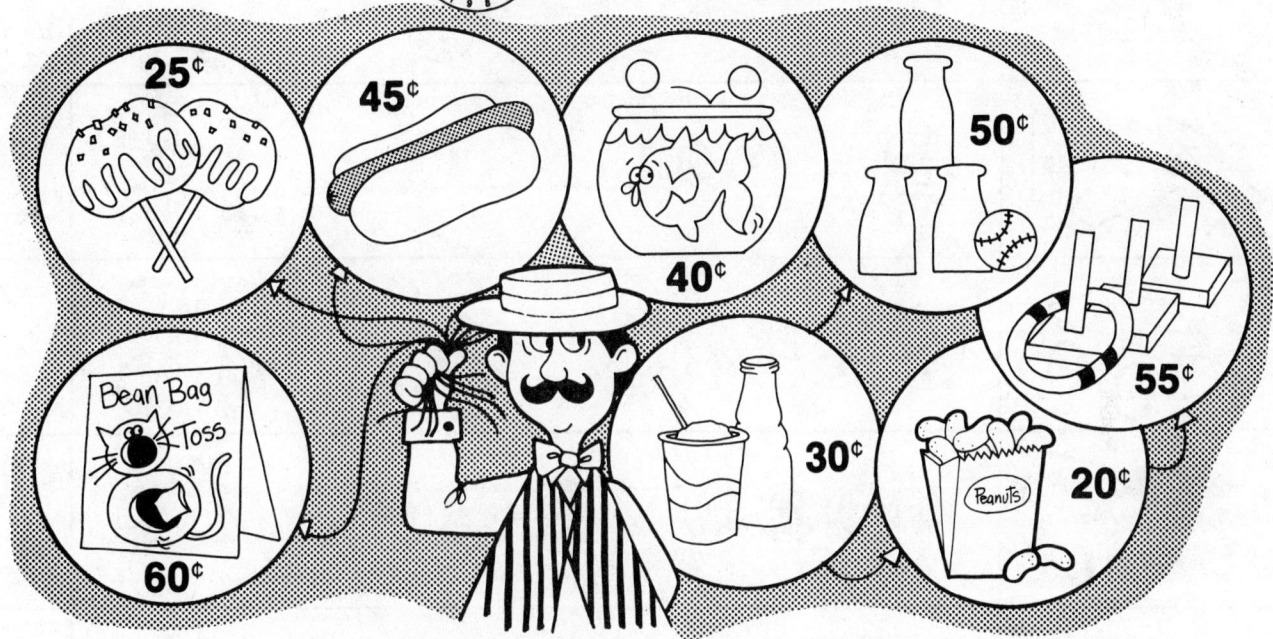

Sunday at the Town Fair
(1) Put an X on the coins you will need to play each game.
(2) How much change is left?
(3) Name a food you can buy with your change.
(The box at the top will help you count.)

		1.	2.	3.
A.	55¢	(half dollar, nickel, dime, nickel)		
B.	50¢	(2 quarters, 6 nickels)		
C.	40¢	(10 dimes)		
D.	60¢	(half dollar, 6 nickels)		

© Frank Schaffer Publications, Inc.   FS-2674 Math Workbook-Time and Money

Name _____ Review: All coins

=50¢ =50¢ =25¢ 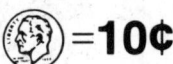=10¢ =5¢ 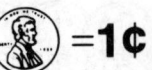=1¢

Put X's on the coins that show how much each toy costs.   Change

1. Ring — 45¢
2. Airplane — 64¢
3. Wagon — 82¢
4. Doll — 50¢
5. Roller skate — 73¢
6. Truck — 61¢
7. Horn — 5¢   BUY 4
8. Kite — 10¢   BUY 5
9. Bubbles — 20¢   BUY 2

Name _____   Skill: Parts of a dollar

| 1. Count by 10  ____ dimes = $1.00 | 2. Count by 25  ____ quarters = $1.00 | 3. Count by 50  ____ half dollars = $1.00 |  $1.00 |

Each grocery order must total $1.00. (1) What coins will you need? Write Q, D, or HD on the line. (2) Write the value of each coin in the box.

4.

| 75¢ | + | 25¢ | = $1.00

Q

7.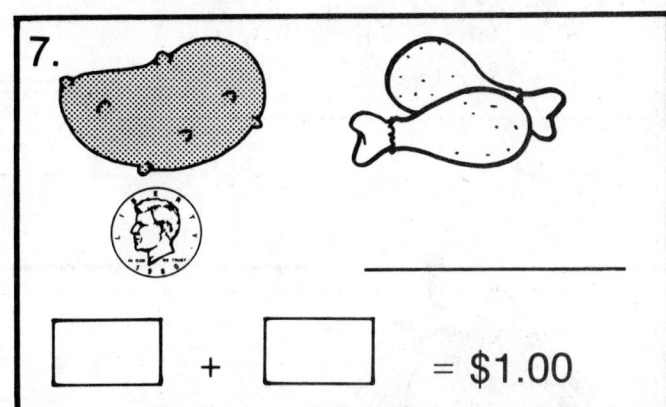

☐ + ☐ = $1.00

5.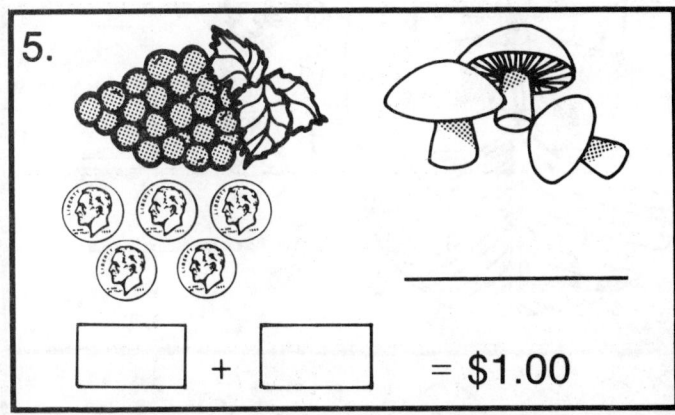

☐ + ☐ = $1.00

8.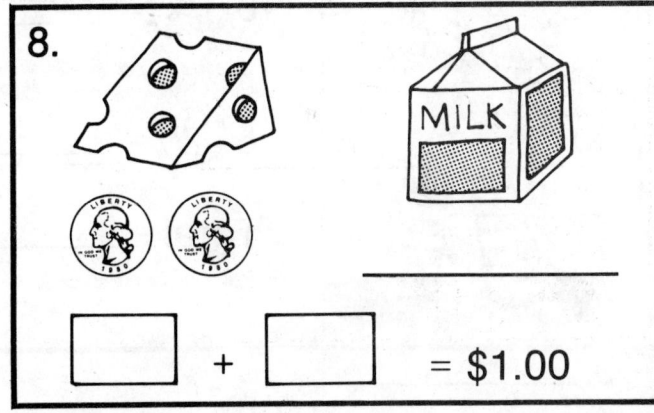

☐ + ☐ = $1.00

6.

☐ + ☐ = $1.00

9.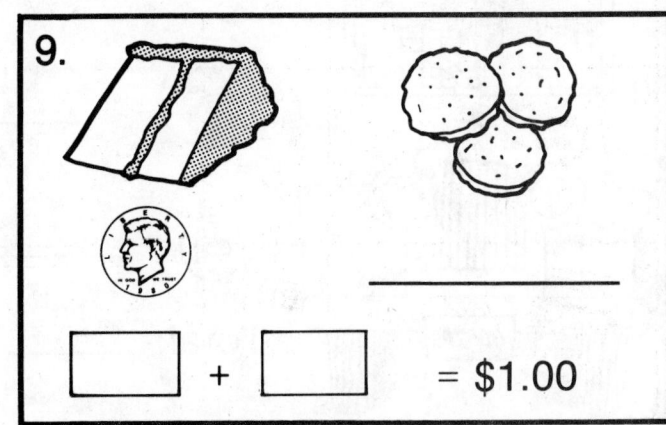

☐ + ☐ = $1.00

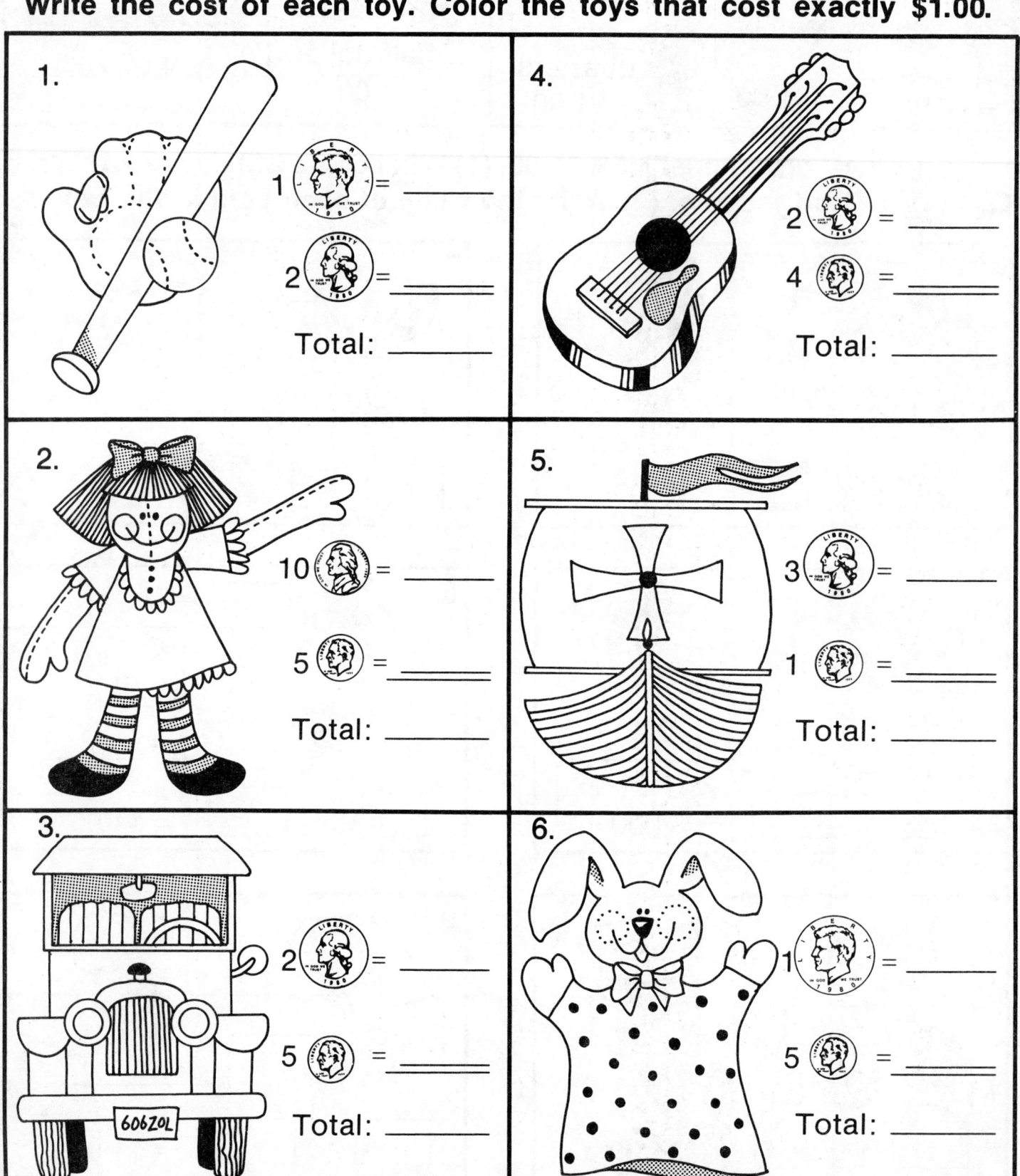

Name _____   Skill: Addition/Subtraction

  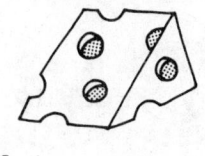

You have   Buy   Cost   How much change do you have left?

1. | $1.55 | 2 MILK = | .40 |
   $ .20

2. | | 6 = | |
   $ .10

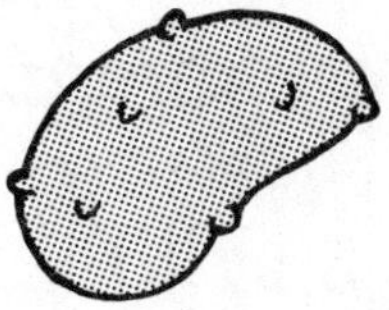

3. | | 3 = | |
   $ .25

4. | | 2 = | |
   $ .75

5. | | 4 = | |
   $ .50

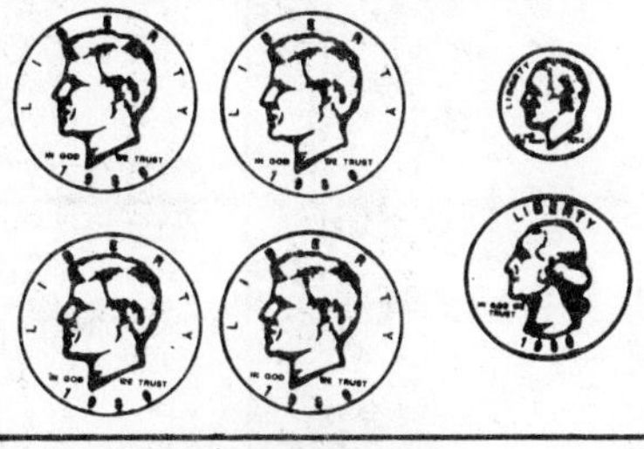

6. | | 10 = | |
   $ .05

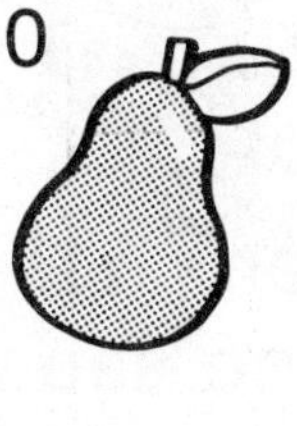

© Frank Schaffer Publications, Inc.   37   FS-2674 MATH WORKBOOK-Time and Money

Name _____ Skill: Addition/Subtraction

seventy-five cents   fifty cents   twenty-five cents   fifteen cents

Cents are often shown with a cents sign (¢) or with a decimal point ($.50). When you have more than a dollar, use a decimal point ($1.55 not 155¢).

Example: Sue has one dollar and twenty-five cents. She wants to buy a 🥧. It costs fifteen cents.

```
   1.25
 -  .15
 $1.10
```

Write the problems. Find the change. (Don't forget the decimal points!)

1. A 💍 costs two dollars and fifty cents. Patrick has one dollar and fifty cents. Subtract to find out how much more money Patrick needs.

2. A 🥤 costs one dollar and forty-five cents. 🍦 is one dollar and twenty cents. Add to find out how much both foods cost.

3. Lou earned two dollars and ten cents. His father gave him one dollar and twenty-five cents. How much did Lou have altogether?

4. Kelly had three dollars and seventy-five cents. She lost one dollar and fifty cents. How much did she have left?

5. Sam's new 🐴 needs a 🪑. The saddle costs four dollars and thirty cents. Sam has four dollars and fifteen cents. Can he buy the saddle? _____ Tell why or why not.

Name _____   Skill: Addition/Subtraction

How much did each girl spend on food? How much change was left?

1. Kate bought peanuts and popcorn.	2. Peggy went to the magic show! Later, she got a soda.	3. Sharon took a pony ride and ate some cookies.
Add / Subtract	Add / Subtract	Add / Subtract

4. Tina bought a soda, cookies and went to the magic show.	5. Who has the most money left over? _____
Add / Subtract	6. Name two things Peggy could buy with her change. _____
	7. Can Kate go to the magic show? _____ Tell why or why not. _____

© Frank Schaffer Publications, Inc.   39   FS-2674 MATH Workbook-Time and Money

Name _____ Skill: Addition/Subtraction

1. Steve has $.40. He bought a toy football for $.30. How much money did he get back?

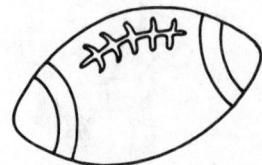

1.

2. Michael earned $.80. He gave his brother $.50. How much did Michael have left?

2.

3. Ellen wants to buy a $.35 ice cream cone. She only has $.15. How much more money does she need?

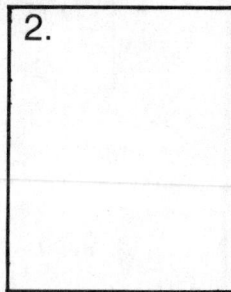

3.

4. Dad gave Tim $1.00. Tim put half the money in the bank. How much did Tim keep?

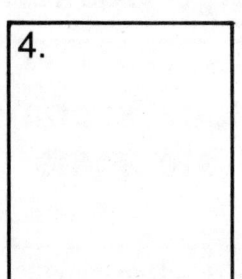

4.

5. A new watch costs $1.85. Randy has $.75. How much more money does Randy need to save?

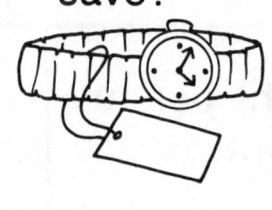

5.

6. A newspaper costs $.20. Barbie gave the man $.90. What change did she get back?

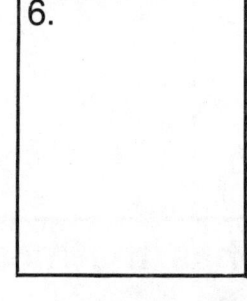

6.

7. John has $.65. He bought two apples. They cost $.15 each. How much did John spend? How much did he have left?

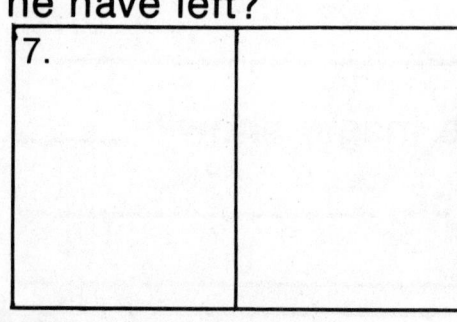

7.

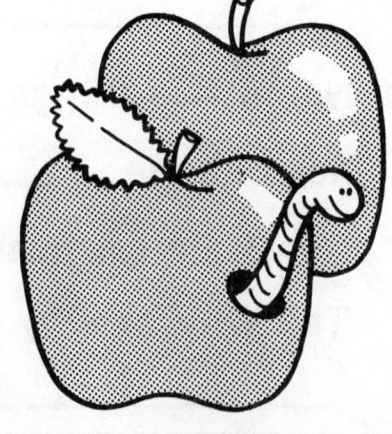

8. Chris bought a sandwich for $.90. She bought an orange for $.25. Chris gave the man $1.15. How much did lunch cost? How much change did Chris get?

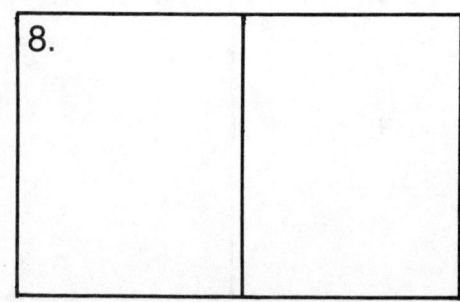

8.

Name _____   Skill: Addition/renaming

How much money did everyone have before they bought ice cream?

1. Mark bought a scoop of banana. His change: $.20

2. Sue bought a scoop of mint. Her change: $.65

3. Pete bought 1 scoop of cherry and 1 scoop of chocolate. His change: $.75

4. Donna bought a scoop of orange. Her change: $.45

5. David bought 2 scoops of vanilla. His change: $.25

6. Linda bought 1 scoop of orange, mint and banana. Her change: $.35

41

Name _____ Skill: Addition/renaming

Diane, Robin, and Ken went to the movies. They each bought a ticket and some food at the snack bar. How much did everyone spend at the movies? How much did each person have left? (Don't forget the decimal points!)

1. Diane has: $2.50

+ _____

Total:

2. Robin has: $4.25

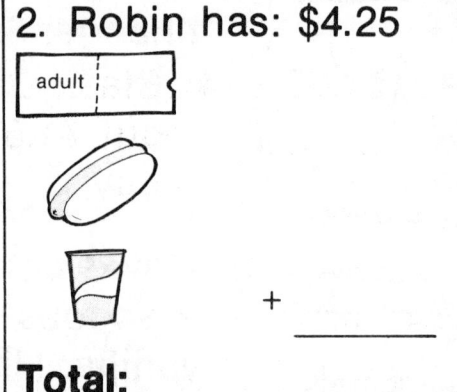

+ _____

Total:

3. Ken has: $4.50

+ _____

Total:

Find out how much each person has left.

| 4. $2.50 | 5. $4.25 | 6. $4.50 |
| − _____ | − _____ | − _____ |

7. Who spent the most money? _____

8. Name two things you can buy that will add up to $1.00. _____
_____

9. Ken wants to buy Diane a hot dog and candy. Will he have enough money? How will you find out? _____

Name _____ Skill: Addition/Subtraction, renaming and regrouping

On Monday, The Dodge City Bank was very busy. Some people took money out. Others put money in. How much money did each person have in the bank at the end of the day?

1. Wild Bill-took out $1.20 to buy a  . in the bank $3.30 _____

4. Black Bart-took out $1.57 to buy a 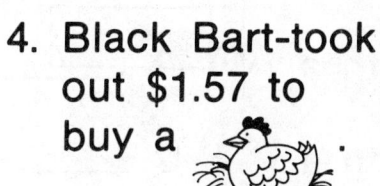 . in the bank $4.65 _____

2. Jess-roped a  . Earned $5.95. in the bank $2.10 _____

5. Minnie-sold a  . Earned $6.48 in the bank $4.20 _____

3. Annie-took out $3.75 for new 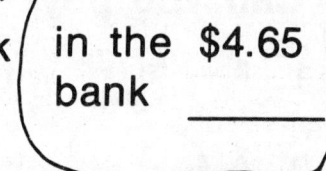 . in the bank $5.65 _____

6. Tiny Tom-took out $8.25 to buy a 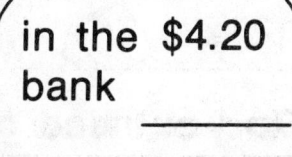 . in the bank $9.54 _____

Make up a word problem of your own. Use these facts: Cowboy Cal wants to buy a horse.

Name _____   Skill: More and less

Can you buy: yes/no   more/less

1. A  with a ? __no__   It costs __more__ than a .

2. A  with a ? _____   It costs _____ than a  .

3. A  with a  ? _____   It costs _____ than a  .

4. A with a ? _____   It costs _____ than a .

Do you need <u>more</u> or <u>less</u> money?

5. You have 🪙🪙 to buy a 🙂 . _____

6. You have 🪙🪙🪙🪙🪙 to buy a 🧁 . _____

7. You have 🪙 to buy a 🍞 . _____

8. You have 🪙🪙 to buy a 🧁 . _____

9. You have 🪙 to buy a ⭐ . _____

10. You have 🪙🪙 to buy a 🥧 . _____

© Frank Schaffer Publications, Inc.   44   FS-2674 Math Workbook-Time and Money

Name _____   Review: Coin recognition and value

1.   2.   3.   4.   5.

Half Dollar
Penny
Quarter
Dime
Nickel

**Write the name of each numbered coin.
Tell its value.**

1. _____ _____ ¢   3. _____ _____ ¢

2. _____ _____ ¢   4. _____ _____ ¢

5. _____ _____ ¢

Change the words into numbers.

6. seventy-five cents _____   10. fifty cents _____

7. thirty cents _____   11. twenty-five cents _____

8. three dollars and
ten cents _____   12. two dollars and
thirty-five cents _____

9. five dollars and
fifteen cents _____   13. one dollar and
twenty-two cents _____

Write the value of these coins.

14.  +  +  +  = _____

15.  + +  +  +  = _____

16. +  +  +  +  = _____

17. + + +  +  = _____

© Frank Schaffer Publications, Inc.   FS-2674 MATH WORKBOOK-Time and Money

Answers

Page One
1. 1:00
2. 5:00
3. 10:00
4. 12:00
5. 6:00
6. 3:00
7. 11:00
8. 8:00
9. 2:00
10. 9:00
11. 3:00
12. 4:00
13. 5:00
14. 7:00

Page Two
a. 7:00
b. 9:00
c. 10:00
d. 4:00
e. 6:00
f. 7:00
g. 9:00
h. 11:00
1. Answers vary.
2. Answers vary.
3. Answers vary.
4. Answers vary.

Page Three
Answers vary.

Page Four
1. 4:00
2. 11:00
3. 7:00
4. 2:00
5. 12:00
6. 11:00
7. 10:00
8. 5:00
9. 10:00
10. 12:00
11. 4:00
12. 12:00

Page Five
1. 8:00 a.m.
2. 5:00 p.m.
3. 2 hours
4. 1:00 p.m.
5. 11 hours
6. 3 hours

Page Six
1. 2:30
2. 6:30
3. 4:30
4. 9:30
5. 11:30
6. 1:30
7. 3:30
8. 10:30
9. 8:30
10. 5:30
11. 7:30
12. 12:30
13. 6:30
14. 1:30

Page Seven
1. 7:30
2. 11:30
3. 1:30
4. 6:30
5. 9:30
6. 4:30
7. 12:30
8. 12:30
9. 9:30
10. 6:30

Page Eight
1. d.
2. b.
3. c.
4. a.
5. 9:30
6. 2:30
7. 6:00
8. 8:30

Page Nine
1. 4:00
2. 3:30
3. 5:00
4. 1:00
5. 9:30
6. 11:00
7. 9:00
8. 10:30
9. 4:30
10. 2:30
11. 6:00
12. 8:30
13. 7:00
14. 12:30

Page Ten
1. 8:00
2. 8:30
3. 9:30
4. 10:00
5. 11:00
6. 11:30
7. 12:00

Page Eleven
1. 1:15
2. 3:15
3. 8:15
4. 10:15
5. 7:15
6. 12:15
7. 9:15
8. 4:15
9. 6:15
10. 11:15
11. 2:15
12. 8:15
13. 10:15
14. 5:15

Page Twelve
1. quarter past eleven
2. quarter past six
3. quarter past five
4. 12:15
5. 5:15
6. 11:15
7. 6:15

Page Thirteen
1. 9:30
2. 2:15
3. 12:30
4. 9:00
5. 11:15
6. 4:00
7. 3:15
8. 8:15
9. 10:30
10. 7:00
11. 1:15
12. 3:30
13. 7:00
14. 12:15

Page Fourteen
1. 3:45
2. 12:45
3. 6:45
4. 9:45
5. 1:45
6. 8:45
7. 4:45
8. 7:45
9. 11:45
10. 5:45
11. 10:45
12. 2:45
13. 6:45
14. 1:45

Page Fifteen
1. 12:45
2. 4:45
3. 10:45
4. 5:45
5. 2:45
6. 8:45
7. 2:45
8. 4:45

Page Sixteen
1. 8:00
2. 8:45
3. 9:30
4. 6:45
5. 2:15
6. 4:45
7. 7:00
8. 1:30
9. 11:15
10. 4:45
11. 9:30
12. 5:15
13. 10:45
14. 6:15

Page Seventeen
1. 1:10
2. 1:20
3. 1:30
4. 1:40
5. 1:50
6. 2:10
7. 2:40
8. 2:50
9. 8:40
10. 3:10
11. 7:20
12. 9:10
13. 5:10
14. 6:50

Page Eighteen
1. 9:40
2. 8:50
3. 11:20
4. 4:10
5. 4:20
6. 10:50
7. 6:40
8. 12:30
9. 30
10. 40
11. 20

Page Nineteen
1. 2:40
2. 2:45
3. 2:50
4. 2:55
5. 12:05
6. 12:15
7. 12:20
8. 12:25
9. 3:25
10. 4:05
11. 9:55
12. 6:35
13. 10:25
14. 10:30

Page Twenty
1. 3:05
2. 10:40
3. 8:50
4. 4:20
5. 7:40
6. 1:15
7. 11:20
8. 7:00
9. 20
10. 30
11. 25

Page Twenty-One
1. 12:00
2. 9:00
3. 8:30
4. 6:00
5. 3:00
6. 2:00
7. 9:30
8. 10:15
9. 11:00
10. 8:00
11. 5:30
12. 7:00
13. 1:45
14. 3:30

Page Twenty-Two
1. 5:10
2. 5:55
3. 2:15
4. 8:05
5. 6:30
6. 9:20
7. 2:45
8. 2:35
9. 6:45
10. 10:50
11. 3:40
12. 1:45
13. 7:05
14. 12:25

Page Twenty-Three
1. 5:40
2. 8:10
3. 1:55
4. 3:25
5. 2:50
6. 11:45
7. 4:05
8. 7:35
9. 10:10
10. 5:55
11. 12:15
12. 3:25
13. 9:05
14. 6:20

Page Twenty-Four
1. 6:00
2. 8:30
3. 5:15
4. 4:45
5. 1:00
6. 9:05
7. 12:35
8. 3:25
9. 11:10
10. 1:20
11. 7:40
12. 5:50
13. 8:55
14. 10:00

Answers

Page Twenty-Five
1. 7:25
2. 8:00
3. 2:05
4. 6:45
5. 3:05
6. 1:20
7. 10:30
8. 12:55
9. 6:40
10. 11:10
11. 5:15
12. 9:40
13. 8:35
14. 4:25

Page Twenty-Six
1. 6:00
2. 3:30
3. 9:15
4. 1:20
5. 7:45
6. 4:30
7. 2:15
8. 6:30
9. 15
10. 35
11. 45
12. 20
13. 2:30
14. 2:15
15. 3:10
16. 8:00

Page Twenty-Seven
1. 3 dimes = 30¢
   4 nickels = 20¢
   Total = 50¢
2. 4 dimes = 40¢
   3 nickels = 15¢
   Total = 55¢
3. 30¢ (Lisa)
4. 25¢ (Terry)
5. 25¢ (Terry)
6. 10¢ (Lisa)
   Lisa's change = 10¢
   Terry's change = 5¢
7. 10¢
8. 15¢
9. 10¢

Page Twenty-Eight
1. 27¢
2. 58¢
3. 35¢
4. 60¢
5. Dimes - 9
   Nickels - 1
   Price - 95¢
6. Dimes - 5
   Nickels - 3
   Price - 65¢
7. Dimes - 7
   Nickels - 4
   Price - 90¢
8. Dimes - 6
   Nickels - 2
   Price - 70¢
9. Dimes - 3
   Nickels - 6
   Price - 60¢

Page Twenty-Nine
1. 5 + 5 + 5 + 5 + 5 + 1 = 26¢
2. 5 + 5 + 5 + 1 + 1 = 17¢
3. 5 + 5 + 5 = 15¢
4. 5 + 5 + 5 + 1 + 1 + 1 = 18¢
5. 5
6. 3
7. 6
8. 9
9. 2
10. 8
11. 8
12. 1
13. 10
14. 6
15. 7
16. 4

Page Thirty
1. 15¢, dime
2. 15¢, 2 nickels
3. 20¢, nickel
4. 10¢, 3 nickels
5. 5¢
6. 10¢
7. 0
8. 15¢
9. yes, 5¢

Page Thirty-One
1. Cost: 50¢, Change: 50¢
2. Cost: 75¢, Change: 50¢
3. Cost: 50¢, Change: 25¢
4. Cost: 50¢, Change: 25¢
5. Cost: 40¢, Change: 10¢
6. Cost: 40¢, Change: 5¢
7. Cost: 25¢, Change: 15¢
8. Cost: 90¢, Change: 30¢
9. Cost: 60¢, Change: 0

Page Thirty-Two
1. 25¢ + 10¢ + 5¢ + 5¢ = 45¢
2. 25¢ + 5¢ + 10¢ + 10¢ + 5¢ = 55¢
3. 25¢ + 10¢ + 10¢ + 5¢ + 5¢ = 55¢
4. 10¢ + 10¢ + 10¢ + 5¢ + 5¢ = 40¢

Page Thirty-Three
1. 50¢
2. 50¢
3. 50¢
4. 50¢
A. 1. Half dollar and nickel
   2. 20¢
   3. peanuts
B. 1. 2 quarters
   2. 25¢
   3. candied apples
C. 1. 4 dimes
   2. 30¢
   3. cola
D. 1. Half dollar and dime
   2. 45¢
   3. hot dog

Page Thirty-Four
1. Answers will vary, 23¢
2. Answers will vary, 30¢
3. Answers will vary, 32¢
4. Answers will vary, 7¢
5. Answers will vary, 11¢
6. Answers will vary, 2¢
7. Answers will vary, 5¢
8. Answers will vary, 22¢
9. Answers will vary, 6¢

Page Thirty-Five
1. 10
2. 4
3. 2
4. Q, 75¢ + 25¢ = $1.00
5. Answers vary, 50¢ + 50¢ = $1.00
6. Q, 75¢ + 25¢ = $1.00
7. Answers vary, 50¢ + 50¢ = $1.00
8. Answers vary, 50¢ + 50¢ = $1.00
9. Answers vary, 50¢ + 50¢ = $1.00

Page Thirty-Six
1. $.50, $.50, Total: $1.00
2. $.50, $.50, Total: $1.00
3. $.50, $.50, Total: $1.00
4. $.50, $.40, Total: $.90
5. $.75, $.10, Total: $.85
6. $.50, $.50, Total: $1.00

Page Thirty-Seven
1. $1.55 − $.40 = $1.15
2. $1.70 − $.60 = $1.10
3. $.95 − $.75 = $.20
4. $2.90 − $1.50 = $1.40
5. $2.35 − $2.00 = $.35
6. $1.95 − $.50 = $1.45

Page Thirty-Eight
1. $2.50
   −1.50
   $1.00
2. $1.45
   +1.20
   $2.65
3. $2.10
   +1.25
   $3.35
4. $3.75
   −1.50
   $2.25
5. $4.30
   −4.15
   $ .15

No, because he needs fifteen more cents.

Answers

Page Thirty-Nine
1. $ .85 $2.85
   + .75   -1.60
   $1.60 $1.25

2. $1.50 $3.00
   + .50   -2.00
   $2.00 $1.00

3. $1.00 $3.75
   + .50   -1.50
   $1.50 $2.25

4. $ .50
     .50 $3.90
   +1.50   -2.50
   $2.50 $1.40

5. Sharon
6. cookies and soda
7. No, The show costs more than $1.25.

Page Forty
1. $ .40
   - .30
   $ .10

2. $ .80
   - .50
   $ .30

3. $ .35
   - .15
   $ .20

4. $1.00
   - .50
   $ .50

5. $1.85
   - .75
   $1.10

6. $ .90
   - .20
   $ .70

7. $ .15 $ .65
   + .15   - .30
   $ .30 $ .35

8. $ .90 $1.15
   + .25   -1.15
   $1.15   0

Page Forty-One
1. $ .30
   + .20
   $ .50

2. $ .20
   + .65
   $ .85

3. $ .35
     .40
   + .75
   $1.50

4. $ .65
   + .45
   $1.10

5. $ .55
     .55
   + .25
   $1.35

6. $ .65
     .20
     .30
   + .35
   $1.50

Page Forty-Two
1. $ .75
     .30
   + .50
   $1.55

2. $1.50
    1.00
   + .70
   $3.20

3. $1.50
     .50
   + .30
   $2.30

4. $2.50
   -1.55
   $ .95

5. $4.25
   -3.20
   $1.05

6. $4.50
   -2.30
   $2.20

7. Robin
8. small drink and popcorn or peanuts and large drink
9. yes, Add $1.00 + $.45.

Page Forty-Three
1. $3.30   4. $4.65
   -1.20   -1.57
   $2.10 $3.08

2. $2.10   5. $4.20
   +5.95   +6.48
   $8.05 $10.68

3. $5.65   6. $9.54
   -3.75   -8.25
   $1.90 $1.29

Page Forty-Four
1. no, more
2. no, more
3. yes, less
4. no, more
5. less
6. more
7. more
8. more
9. less
10. more

Page Forty-Five
1. Half dollar, 50¢
2. Quarter, 25¢
3. Nickel, 5¢
4. Penny, 1¢
5. Dime, 10¢
6. $.75
7. $.30
8. $3.10
9. $5.15
10. $.50
11. $.25
12. $2.35
13. $1.22
14. 45¢
15. 36¢
16. 25¢
17. 23¢